Methodology for Biomimetic Design of Additively Manufactured Structural Components

Vom Promotionsausschuss der

Technischen Universität Hamburg

zur Erlangung des akademischen Grades

Doktor-Ingenieur (Dr.-Ing.)

genehmigte Dissertation (kumulativ)

von

Tim Röver

aus Bad Segeberg

2026

1. Gutachter: Prof. Dr.-Ing. Claus Emmelmann

2. Gutachter: Prof. Dr.-Ing. Otto von Estorff

Tag der mündlichen Prüfung: 29.10.2025

Light Engineering für die Praxis

Reihe herausgegeben von

Claus Emmelmann, Laser Zentrum Nord GmbH, Hamburg, Germany

Technologie- und Wissenstransfer für die photonische Industrie ist der Inhalt dieser Buchreihe. Der Herausgeber leitet das Institut für Laser- und Anlagensystemtechnik (iLAS) an der Technischen Universität Hamburg (TUHH). Die Inhalte eröffnen den Lesern in der Forschung und in Unternehmen die Möglichkeit, innovative Produkte und Prozesse zu erkennen und so ihre Wettbewerbsfähigkeit nachhaltig zu stärken. Die Kenntnisse dienen der Weiterbildung von Ingenieuren und Multiplikatoren für die Produktentwicklung sowie die Produktions- und Lasertechnik, sie beinhalten die Entwicklung lasergestützter Produktionstechnologien und der Qualitätssicherung von Laserprozessen und Anlagen sowie Anleitungen für Beratungs- und Ausbildungsdienstleistungen für die Industrie.

Tim Röver

Methodology for Biomimetic Design of Additively Manufactured Structural Components

Springer

Tim Röver
Hamburg University of Technology
Hamburg, Hamburg, Germany

ISSN 2522-8447 ISSN 2522-8455 (electronic)
Light Engineering für die Praxis
ISBN 978-3-032-27404-5 ISBN 978-3-032-27405-2 (eBook)
https://doi.org/10.1007/978-3-032-27405-2

This work was supported by Technische Universität Hamburg.

This Springer imprint is published by the registered company Springer Nature Switzerland AG
The registered company address is: Gewerbestrasse 11, 6330 Cham, Switzerland

Preface

This work was developed during my time as a research associate and research group leader at the Institute of Laser and System Technologies (iLAS) at Hamburg University of Technology (TUHH).

First and foremost, I would like to sincerely thank my doctoral advisor, Professor Dr.-Ing. Claus Emmelmann, for his great trust and the constructive discussions, which often resulted in a better overall analysis and a clearer focus on the bigger picture. I would also like to warmly thank Professor Dr.-Ing. Otto von Estorff for serving as the second reviewer and Professor Dr.-Ing. Jan Dege for chairing the examination committee.

Furthermore, I wish to express my gratitude to all colleagues at iLAS for the friendly atmosphere and support. I would like to make special mention of Karim Asami for the many valuable discussions and ideas. My heartfelt thanks also go to all co-authors of the four scientific articles that have been published as part of this dissertation.

I would also like to acknowledge the support of German society, which has financed the education system and, indirectly, the positions at TUHH from which I have benefited and which have made it possible for me to write this book.

I am deeply grateful to my family and friends, especially my parents, who have placed great value on my education and supported me throughout. Finally, I would like to sincerely thank my life partner, Vivian Soares Arcanjo, for her steady support, encouragement, and patience throughout the entire thesis project.

Hamburg, February 2026
Tim Röver

Abstract

Previous research indicates that incorporating biomimetic beam-like structures into topology optimization designs can enhance the lightweight characteristics of components such as aviation brackets that were produced by the additive manufacturing technique of powder bed fusion of metal with a laser beam. However, there is a need for detailed design methodologies to effectively implement such design modifications. The first research question of this thesis explores how a design methodology can be developed to generate biomimetic design components from density-based topology optimization designs. The secondary research question assesses the lightweight characteristics of the outcomes generated by such a methodology, comparing them to those of the input topology optimization design.

A five-step design methodology with detailed submethodologies was developed. It takes a topology optimization design as an input and, in the first step generates an auxiliary 3D model consisting of cylindrical beams and spherical nodes that resembles the topology optimization design. A skeletonization approach is used for abstraction. In the second step, a finite element analysis of the auxiliary model is performed to evaluate the internal forces and moments in its beams. In step three, two biomimetic parametric beam designs as well as two conventional parametric beam designs are considered for parameter optimization for each of the beam load cases from the auxiliary model. The most lightweight optimized beam is selected for each of the beams of the abstraction, respectively. Buckling analyses and re-dimensioning of optimized beams are performed, if necessary. In step four, adapted nodes for the biomimetic component design are generated. The nodes contain a lattice infill structure to ensure powder-removability and a lightweight design. In step five, the complete biomimetic component design is validated by finite element analysis and the mass of the component is evaluated. Thus, a complete design methodology was developed successfully.

Three common topology optimization problem formulations were considered for validation of the developed biomimetic design methodology. Finite element analyses of the biomimetic designs yielded the existence of critical regions of high stress in all three of the biomimetic components. However, the locations of critical stresses have been associated with specific regions within the biomimetic designs. Mass evaluation showed that the biomimetic component designs are 12.5 % - 30.3 % lighter compared to their topology optimization counterparts.

The developed methodology stands out from previously existing research as it considers the internal forces and moments in beams and closes a research gap for detailed design methodologies for biomimetic structural components. This thesis contributes to harnessing the immense potential of biomimetic design in lightweight industries, such as aerospace and automotive. Future research should refine the presented methodology and perform experimental assessment of component designs generated by it.

Table of contents

List of figures

List of tables

List of abbreviations

Abbreviation	Meaning
1D	One-dimensional
2D	Two-dimensional
3D	Three-dimensional
Al-Si-Mg	Aluminum alloy that primarily consists of aluminum (Al) as well as silicon (Si), and magnesium (Mg)
AlSi10Mg	Aluminum alloy that primarily consists of aluminum (Al), approximately 10 % silicon (Si), and a small percentage of magnesium (Mg)
AM	Additive manufacturing
ASTM	ASTM International, formerly known as American Society for Testing and Materials
BJT	Binder jetting
CAD	Computer-aided design
CO_2	Carbon dioxide
COMSOL 5.4	COMSOL Multiphysics 5.4
COMSOL 6.0	COMSOL Multiphysics 6.0
DED	Directed energy deposition
DfAM	Design for additive manufacturing
DIN	Deutsches Institut für Normung e.V. (German Institute for Standardization)
EliSE	Evolutionary light structure engineering
EN	Europäische Norm (European Standard)
FCRC	Flight crew rest compartment
FEA	Finite element analysis
FEM	Finite element method
HC	Parametric hollow cylindrical beam
IN718	Specific nickel-based superalloy
Inventor 2024	Autodesk Inventor 2024
ISO	International Organization for Standardization
L-PBF	Laser powder bed fusion
MBB	Messerschmitt-Bölkow-Blohm
MEX	Material extrusion
Netfabb	Autodesk Netfabb Ultimate slicer software (version 2023.1)
nTop	nTop 4.22.2
PBF	Powder bed fusion
PBF-EB/M	Powder bed fusion of metal with electron beam
PBF-LB	Powder bed fusion with a laser beam
PBF-LB/M	Powder bed fusion of metal with a laser beam
PBF-LB/P	Powder bed fusion of polymer with a laser beam
Q1	Primary research question
Q2	Secondary research question
RD1	Parametric revolver drum beam 1
RD2	Parametric revolver drum beam 2
SC	Parametric solid cylindrical beam
SHL	Sheet lamination
SLM 800	Name of a specific PBF-LB/M machine

Abbreviation	Meaning
SIMP	Simplified isotropic material with penalization
STEP	Standard for the exchange of product model data
STL	Stereolithography
Ti-6Al-4V	Titanium alloy that primarily consists of titanium (Ti), contains 6 % aluminum (Al), and 4 % vanadium (V)
UAV	Unmanned aerial vehicle
VPP	Vat photopolymerization

List of symbols

Symbol	Meaning	Unit
E	Young's modulus of the material	Pa
E_0	Young's modulus of the fully dense material in power-law	Pa
E_{AlSi10Mg}	Young's modulus of aluminum alloy AlSi10Mg equal to 57 GPa	Pa
E_{eff}	Effective Young's modulus in power-law	Pa
F	Force	N
F_{TO}	Objective function in topology optimization problem	-
$\mathbf{F}$	Force vector	N
F_{crit}	Critical force of beam under buckling	N
G_0	Volume constraint function in topology optimization problem	-
G_j	j-th additional constraint function in topology optimization problem	-
I	Area moment of inertia of the cross section about the axis of bending	m^4
M_1	Internal bending moment around the first local axis in beam cut-surface	N×m
$M_{1,\mathrm{PO}}$	Bending moment around the first local axis load for parameter optimization	N×m
M_2	Internal bending moment around the second local axis in beam cut-surface	N×m
$M_{2,\mathrm{PO}}$	Bending moment around the second local axis load for parameter optimization	N×m
M_{t}	Internal twisting moment in beam cut-surface	N×m
$M_{\mathrm{t,PO}}$	Twisting moment input load for parameter optimization	N×m
$M_{\mathrm{xy},20}$	Internal xy-bending moment component at 20 % of length of beam	N×m
$M_{\mathrm{xy},80}$	Internal xy-bending moment component at 80 % of length of beam	N×m
$M_{\mathrm{xz},20}$	Internal xz-bending moment component at 20 % of length of beam	N×m
$M_{\mathrm{xz},80}$	Internal xz-bending moment component at 80 % of length of beam	N×m
N_{FE}	Number of finite elements in topology optimization problem	-
N	Internal normal force in beam cut-surface	N
N_{PO}	Axial force input load for parameter optimization	N
$N_{\mathrm{x},20}$	Internal normal force at 20 % of length of beam	N
$N_{\mathrm{x},80}$	Internal normal force at 80 % of length of beam	N
$S_{\mathrm{xy},20}$	Internal xy-shear force component at 20 % of length of beam	N
$S_{\mathrm{xy},80}$	Internal xy-shear force component at 80 % of length of beam	N
$S_{\mathrm{xz},20}$	Internal xz-shear force component at 20 % of length of beam	N

Symbol	Meaning	Unit
$S_{xz,80}$	Internal xz-shear force component at 80 % of length of beam	N
T_1	Internal shear force along the first local axis in beam cut-surface	N
$T_{1,PO}$	Shear force along the first local axis input load for parameter optimization	N
T_2	Internal shear force along the second local axis in beam cut-surface	N
$T_{2,PO}$	Shear force along the second local axis input load for parameter optimization	N
$T_{x,20}$	Internal torque at 20 % of length of beam	N×m
$T_{x,80}$	Internal torque at 80 % of length of beam	N×m
V_0	Total allowed volume of material in the design domain in topology optimization problem	m^3
dV	Differential volume in topology optimization problem	m^3
f	Local function in topology optimization problem	-
g	Interpolation function in power law	-
l	Length of component	m
p	Penalization parameter in power-law	-
r	Radius	m
r_i	Inner radius	m
r_{k1}	Radius of cavities in first cavity ring	m
r_{k2}	Radius of cavities in second cavity ring	m
t_1	Wall thickness of first ring	m
t_2	Wall thickness of second ring	m
t_3	Wall thickness of third ring	m
t_{kmin1}	Minimum distance between cavities in first cavity ring	m
t_{kmin2}	Minimum distance between cavities in second cavity ring	m
$\mathbf{u}$	State field in topology optimization problem	-
v_i	Volume of the *i*-th element in topology optimization problem	m^3
x	x-coordinate; displayed in non-italic font in figures due to software limitations	m
y	y-coordinate; displayed in non-italic font in figures due to software limitations	m
z	z-coordinate; displayed in non-italic font in figures due to software limitations	m
Ω_i	Volume of the *i*-th element	m^3
$\nu_{AlSi10Mg}$	Poisson's ratio of aluminum alloy AlSi10Mg equal to 0.34	-
π	Mathematical constant representing the ratio of a circle's circumference to its diameter approximately equal to 3.14159	-
$\boldsymbol{\rho}$	Design variable vector	-
$\rho_{AlSi10Mg}$	Density of aluminum alloy AlSi10Mg equal to 2.67 g/cm^3	kg/m^3
ρ_i	*i*-th density variable	-
$\sigma_{acc,AlSi10Mg}$	Maximum acceptable stress value for aluminum alloy AlSi10Mg equal to 112.8 MPa	Pa

Symbol	Meaning	Unit
$\sigma_{y,AlSi10Mg}$	Yield strength of aluminum alloy AlSi10Mg equal to 141 MPa	Pa

Declaration of artificial-intelligence-assisted technologies in the writing process

In the preparation of this work, the author used TUKI (https://openai.rz.tuhh.de/login) to enhance the language quality. After employing this tool, the author reviewed and edited the content as needed and takes full responsibility for the final content of this thesis.

1. Introduction

Time, cost, and quality are important factors in project management [1] and product development processes. These factors are closely related to the well-known *Iron Triangle* first described by Dr. Martin Barnes [2]. In addition to these well-established drivers, sustainability has become very important in recent years, and is considered by some a fourth dimension [3]. In its efforts to tackle climate change, the European Union formulated the goal to use sustainable mobility and transport systems [4]. To enhance the sustainability of mobility systems, a key strategy is to minimize the amount of material used in their components by adopting lightweight design [5]. Another strategy is the use of lightweighting to reduce fuel consumption and emissions during service of mobility systems such as aircraft [6] as e.g. carbon dioxide (CO_2) emissions are considered a great driver of climate change [7]. Due to this, lightweight engineering remains of great importance to achieve sustainable technical solutions. One important tool in lightweight design is the use of *topology optimization* to find highly efficient design concepts. Topology optimization has been used for many years now in the fields of automotive [8] and aerospace [9]. Designs generated with the aid of topology optimization are often geometrically complex and appear organic in shape. They often are numerical approximations of mathematical optima based on the formulated problem description of physical systems.

Additionally, component designs from topology optimization can be improved with regard to lightweight by consideration of biomimetic design principles. A study showed that a new component designs (see Figure 1.1) based on topology optimization, biomimetics, additive manufacturing (AM) and a change of material resulted in weight reduction of 50 % for an aerospace bracket [10]. The authors state that incorporation of bamboo-inspired beams are one of the reasons for the great lightweight character of the developed component. A scientific review article from 2020 focuses on the symbiosis of biomimetic design and laser-based AM [11]. The authors review and discuss the added value of biomimetics in laser-based AM and conclude that the two fields can benefit from each other. Furthermore, they identify the lack of biomimetic design tools as one major obstacle for the wide use of biomimetic designs in combination with laser-based AM [11].

This thesis presents a design methodology for generating detailed and well-dimensioned designs of biomimetic components based on topology optimization results. Therefore, the thesis contributes to closing the gap between AM and biomimetic design to allow for a wider use of biomimetic structural designs in research and industry.

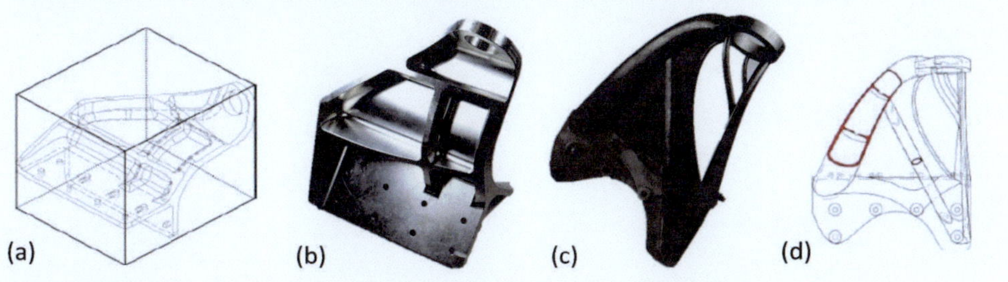

Figure 1.1. Bracket fixing flight crew rest compartment (FCRC) with the primary aircraft hull. (a)-(b): FCRC bracket in conventional design; (c)-(d): FCRC bracket in biomimetic design for additive manufacturing, with the bamboo-like structured section colored in red. Modified and reproduced from [11] (licensed under CC BY 4.0).

T. Röver, *Methodology for Biomimetic Design of Additively Manufactured Structural Components*, Light Engineering für die Praxis, https://doi.org/10.1007/978-3-032-27405-2_1

Industries and applications that could benefit from the methodology are any truss-like mechanical components. Potential applications in aerospace are satellite antenna brackets [12–14], satellite reaction wheel brackets [15], civil aviation brackets [16,17], and space rocket launcher brackets [18]. In automotive the methodology could improve motorcycle vehicle structures [19] and suspension and vehicle structure components in sports cars [20,21]. Furthermore, additively manufactured frames and forks of bicycles [22] could benefit from redesign using the methodology.

This thesis is structured into five chapters, namely, 1. Introduction, 2. Fundamentals of biomimetic design for AM, 3. Research questions, methodology, and materials, 4. Biomimetic design methodology, and 5. Conclusion and outlook. In this introduction the motivation behind this thesis was stated and a gap in research was introduced. In chapter 2, literature and information that is of high relevance for this thesis is introduced. In chapter 3, the research question is defined and the general methodology to address the research questions is presented. Chapter 4 contains the main part of this thesis in which the design methodologies that were developed within this thesis and the results that were generated are presented. In chapter 5, the conclusion as well as an outlook is given. In 0. References, all literature referenced in this thesis can be found. As this thesis is a cumulative dissertation large parts of the contents of this thesis, especially those of chapter 4, can also be found in the respective publications [23–26].

2. Fundamentals of biomimetic design for AM

In the following sections, relevant fundamentals with regard to biomimetics and design for additive manufacturing (DfAM) are presented.

2.1 Additive manufacturing

The terminology and fundamentals of AM are defined in DIN EN ISO/ASTM 52900 [27]. AM is the general term for technologies in which materials are successively joined together to produce physical objects according to the specifications of three-dimensional (3D) model data. Numerous AM processes exist and they can be categorized. Examples of these categories are binder jetting, directed energy deposition, material extrusion and powder bed fusion [27]. Each of the processes may have additional process characteristics which allow for a more detailed terminology of the respective process. An example is given by *powder bed fusion of metal with a laser beam* which may be abbreviated with PBF-LB/M.

2.1.1 Powder bed fusion of metal with a laser beam

In Figure 2.1 a schematic of a PBF-LB/M process is given. The process can be broken down into five steps: pre-processing, recoating, exposure, lowering platform, and post-processing [28]. In pre-processing, the 3D model of the part is converted into a compatible file format, typically STL (originating from the term stereolithography). The model is sliced into thin layers, usually 20-100 μm thick [29], and laser paths based on various possible scanning strategies are converted into a file that can be read by the employed powder bed fusion (PBF) machine. The PBF machine is then calibrated.

Then, a recoater, which usually is a blade, spreads a thin, uniformly distributed layer of powder onto the building platform. Following this, the exposure step involves the laser selectively melting powder. As the laser moves, the melted powder cools and solidifies, forming a solid layer. Once a layer is fused, the build platform is lowered by one layer thickness. Then the recoater applies another layer of powder to start another iteration of the manufacturing process. These steps are repeated for each layer, gradually building up the part layer by layer. [30]

After the build is complete, several post-processing steps are necessary to finalize the part. The build platform with the finished part is removed from the PBF machine, and excess and unfused powder is brushed or blasted away, often for reuse in future builds [31]. The part may undergo heat treatment and hot isostatic pressing to relieve internal stresses induced during the laser melting process and to tune the material properties [30]. Next, the separation of the part from the build platform is performed by, e.g., electrical discharge machining or a saw. Additional finishing processes such as machining, polishing, or coating may be applied to achieve the desired surface finish and dimensional accuracy. For more detailed information on this matter the reader is kindly referred to one of the standard books in the field of AM by Gibson et al. [31] and an overview article by Herzog et al. [30].

T. Röver, *Methodology for Biomimetic Design of Additively Manufactured Structural Components*, Light Engineering für die Praxis, https://doi.org/10.1007/978-3-032-27405-2_2

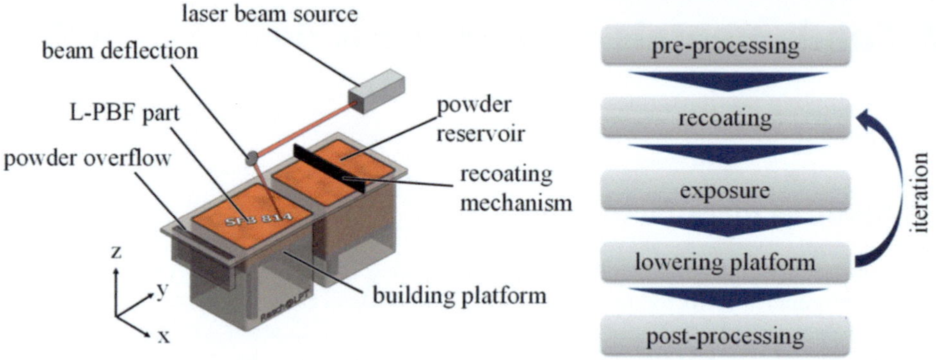

Figure 2.1. Schematic of the PBF-LB/M process. Reproduced from [28] (licensed under CC BY 4.0).

The first inventions in the field of PBF processes date back to the years 1971 and 1979 and in 1995 a project consortium around EOS GmbH from Germany developed parameters for laser-based AM of dense metallic parts in a powder bed process [32]. Since then the metal AM market has grown rapidly with a market value of 3.47 billion in 2023 and it is expected to have a compound annual growth rate of 20.4 % until 2028 [33]. It is noted that the metal AM market is expected to grow twice as fast as the polymer AM market until 2028 [33]. Additionally, metal powder bed fusion alone takes 40 % of revenue share for equipment sales considering all polymer and metal technologies [33]. This underlines the importance of this technology. Furthermore, powder bed fusion with a laser beam (PBF-LB) has the greatest technological maturity as well as greatest relevance in industry considering all AM technologies [33]. Applications for metal AM can be found in aerospace, biomedical, automotive, dental and other fields [34]. In a 2019 industry report [35] DfAM was identified as a critical topic for the success of AM in production applications and also a more recent industry report [33] considers DfAM a focal point in metal AM. Additionally, Leary states that the opportunities of AM depend on the availability of DfAM tools [36].

2.1.2 Design for AM

DfAM refers to the development, design and dimensioning of components considering the technical process constraints of the respective AM process while leveraging the potential given by the design freedom of the process in comparison to conventional manufacturing processes to achieve highly efficient components. To identify the potential given by AM processes, research was done on the development of design guidelines for e.g. the PFB-LB/M process for the materials titanium alloy Ti-6Al-4V [37] and nickel-based superalloy IN718 [38]. In both studies, the manufacturability of geometrical features such as thin walls, pins, overhanging structures, bore holes (cylindrical voids in AM) were investigated. Based on the results, design guidelines for specific geometrical features were developed that give general recommendations for the design of parts to be manufactured in the respective material by PBF-LB/M [37,38].

Additionally, DIN EN ISO/ASTM 52910 [39] provides information about component design considering AM processes. DIN EN ISO/ASTM 52911-1 [40] complements the previously mentioned standard and defines the characteristics of PBF-LB/M and provides detailed design recommendations. It is stated that in PBF-LB/M, the threshold for the use of support structures is usually a downskin angle of the surface between 30 ° and 45 ° [40].

A study [41] investigated unsupported overhanging surfaces with samples that were manufactured from aluminum alloy AlSi10Mg. The results show satisfactory geometric accuracy for unsupported overhanging surfaces with an overhang angle of 45 ° [41]. Additionally, a specific constraint relevant to this thesis was investigated by Han et al. [42]. In their study, the authors conclude that part support is needed for bore holes whose main axis is orthogonal to the build direction if their nominal diameter is greater than 15 mm. The study focused on aluminum alloy AlSi10Mg [42].

Specific topics in the field of DfAM that are relevant to this thesis are lattice structures and triply periodic minimal surfaces [36]. Many AM technologies are able to produce these kinds of structures whereas they often cannot be produced using conventional manufacturing techniques [36]. Furthermore, they have an overlap with the fields of biomimetics [43] and topology optimization [44], two fields of research that will be briefly introduced in the upcoming sections. Lattice structures allow for low-density metallic components of high strength which is of interest in many applications [43]. Figure 2.2 shows an exemplary lattice structure, namely a Kelvin cell. For detailed information on the use of lattice structures and TPMS in combination with AM the reader is kindly referred to a standard book [36] and relevant articles [45,46] that provide an overview over the field. Utilizing the design freedom of AM processes, topology optimization was also found to be valuable in DfAM for development of efficient component designs.

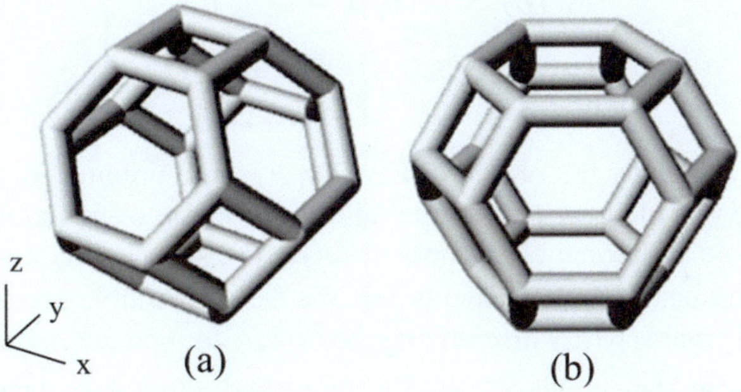

Figure 2.2. The unit cell of Kelvin structures. (a): Kelvin structure with tetragon as base; (b): Kelvin structure with hexagon as base. Modified and reproduced from [47] (licensed under CC BY 4.0).

2.2 Topology optimization

Modern structural optimization is largely based on the seminal paper by Schmit from 1960 [48,49]. Size optimization, shape optimization, and topology optimization are three different approaches that can be used for efficient use of material in component design [50]. Size optimization is also referred to as parameter optimization in this work. Fundamental information on the topic can be found in one of the central papers of this thesis [25]. For additional fundamentals on parameter optimization the reader is kindly referred to the books [48] and [51].

Coming back to topology optimization, it is the most radical approach of the three modern structural optimization approaches, allowing for the generation of a new topology. According to Bendsøe and Sigmund [50] the "purpose of topology optimization is to find the optimal layout of a structure within a specified region". Even though the first works on topology optimization focused on mechanical design problems, topology optimization by now was applied in many fields such as fluids, acoustics, electromagnetics and optics [52].

The works by Bendsøe and Kikuchi from 1988 [53] and Bendsøe from 1989 [54] on the *material distribution method* are important works in the field of topology optimization and may be considered starting points for the development of application-oriented topology optimization approaches. Since then the field of topology optimization has developed rapidly and various approaches such as density, level set, topological derivative, phase field and evolutionary were developed [52]. In the following, the density-based Simplified Isotropic Material with Penalization (SIMP) approach [50] is presented in more detail as it is widely used in commercial software such as COMSOL Multiphysics 6.0 [55], SOLIDWORKS 2023 [56] and Simcenter Nastran Topology Optimization [57], which shows its importance for industry.

2.2.1 The SIMP approach

According to [52] the continuous topology optimization problem can be written as

$$\left.\begin{array}{lrl} \min\limits_{\rho\rho}: & F_{\mathrm{TO}}(\mathbf{u}(\boldsymbol{\rho}),\boldsymbol{\rho}) = \sum_i \int_{\Omega_i} f(\mathbf{u}(\rho_i),\rho_i)\mathrm{d}V & \\ \text{s.t. :} & G_0(\boldsymbol{\rho}) = \sum_i v_i\,\rho_i - V_0 \le 0 & \\ : & G_j(\mathbf{u}(\boldsymbol{\rho}),\boldsymbol{\rho}) \le 0, & j = 1,\dots,M \\ : & 0 \le \rho_i \le 1, & i = 1,\dots,N_{\mathrm{FE}} \end{array}\right\}. \tag{2.1}$$

The variables are given by the objective function F_{TO}, the volume constraint $G_0(\boldsymbol{\rho}) \le 0$, the volume of the i-th element v_i, the total allowed volume of material in the design domain V_0, M possible additional constraints $G_j(\mathbf{u}(\boldsymbol{\rho}),\boldsymbol{\rho}) \le 0,\ j = 1,\dots,M$, the i-th density variable ρ_i, the number of finite elements N_{FE}, the design variable vector $\boldsymbol{\rho}$, the volume of the i-th element Ω_i that is partly discretizing the design domain, the state field $\mathbf{u}$ (satisfying a linear or non-linear state equation) and the simplified integral $\int_{\Omega_i} f(\mathbf{u}(\rho_i),\rho_i)dV$ of the local function $f(\mathbf{u}(\rho_i),\rho_i)$ [52].

Furthermore, according to [52] the power-law that describes the relation between the density design variable and the material property as utilized in the SIMP approach is given by

$$E_{\mathrm{eff}}(\rho_i) = g(\rho_i)E_0 = \rho_i^p E_0,\ \ g(\rho_i) = \rho_i^p\ . \tag{2.2}$$

The variables are given by the effective Young's modulus of the i-th element $E_{eff}(\rho_i)$, the penalization parameter p, and the Young's modulus of the fully dense material E_0 [52]. This approach has proven to be efficient and became popular [50]. Depending on the choice of the penalization parameter the optimization is encouraged moderately (low p-values) or strongly (high p-values) to generate a 0-1 solution. A 0-1 solution is a solution in which each element is either fully dense (ρ_i=1) or is void (ρ_i=0). 0-1 solutions are desirable as they are close to a physically meaningful part design. For more detailed information on the topic the reader is kindly referred to [50].

Figure 2.3 shows a topology optimization procedure that is based on [58] and [50] and is briefly presented in this paragraph. The design space is discretized using finite elements, and mechanical boundary conditions are introduced to the discrete model. Subsequently, a finite element analysis (FEA) is conducted. An optimization algorithm is then employed to modify the element densities. The convergence of the solution is evaluated based on a predefined criterion (e.g. small improvements in compliance with regard to previous iteration). If the solution has not converged, a feedback loop is run to the previous step of FEA, initiating a new iteration. Conversely, if the solution has converged, postprocessing is performed. During postprocessing, the final topology optimization result is assessed using the design variable ρ_i and a threshold (e.g., ρ_i<0.6) to determine the volumetric representation of a part design. Following this evaluation, the procedure is concluded. Optionally, filters based on Helmholtz-type differential equations may be used to improve the topology optimization result [59,60].

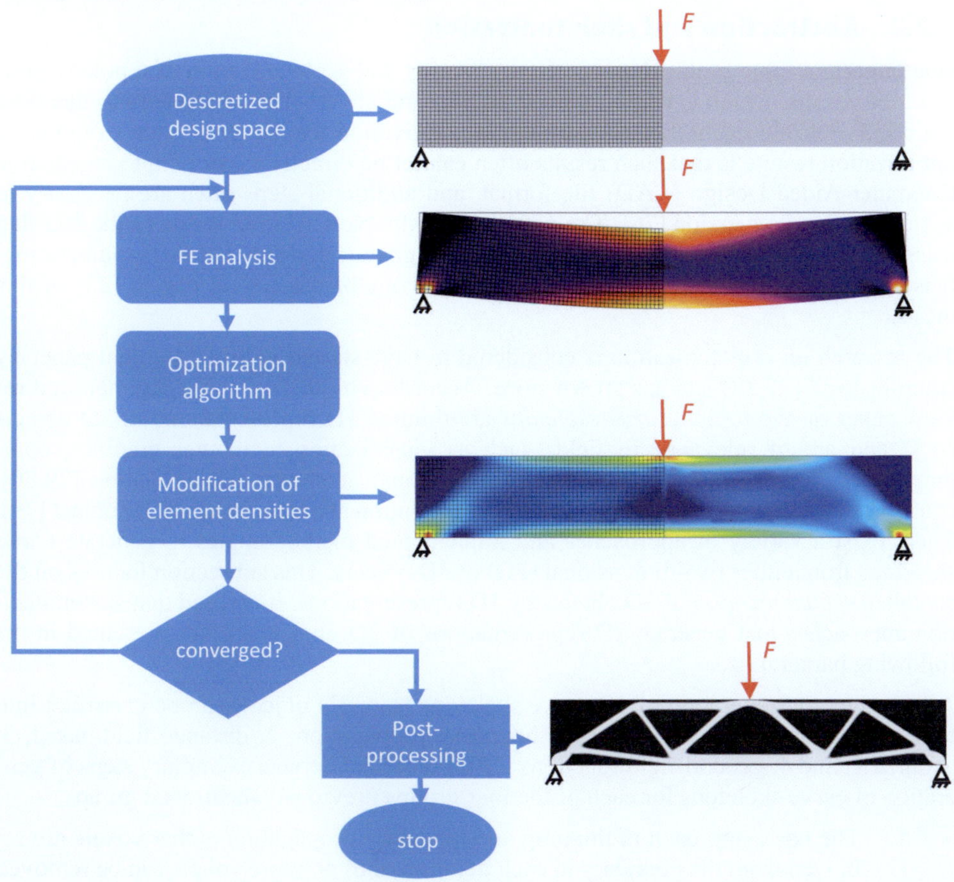

Figure 2.3. Topology optimization procedure partly based on [50,58,61].

Topology optimization has wide use in applications connected to structural design such as structural members of the suspension assembly in motor sports [62], optimized pressure distribution in fuel cell end plates [63], aircraft and aerospace structures [9]. Other fields of application of topology optimization are heat transfer problems [64–69], medical applications [70], electromagnetics [71] and fluid flow [72].

In general topology optimization is often implemented with the help of the finite element method (FEM) that FEA is based on. It is considered that the reader is familiar with FEM and FEA. For information on the topic the reader is kindly referred to the two books [73] and [74]. Coming back to the matter of topology optimization, one important aspect in density-based topology optimization is mesh-dependency. Usually solutions will not converge with mesh-refinement but finer and finer structures will develop when mesh-refinement is done [50]. This is an important limitation that has to be considered when utilizing topology optimization.

In many cases, topology optimization results improve when a larger number of finite elements is used during the optimization process. Even one of the far developed SIMP based formulations that uses a projection scheme shows that the topology of the optimization result converges with mesh refinement [60,75].

2.2.2 Abstraction and skeletonization

One important topic in this thesis is the abstraction and skeletonization of topology optimization results, which is why important aspects and approaches with regard to this field are briefly introduced in this subsection. One motivation for the abstraction of topology optimization results is that such results often cannot be directly exported into an editable Computer-Aided Design (CAD) file format, and additional steps often are necessary to achieve such a conversion [76]. The need for an editable CAD file is due to the fact that usually manual modifications to the generated design are necessary before manufacturing. It is noted that this subsection is partly based on one of the central papers [23] of this thesis.

The research on skeletonization is considered to have started with the seminal paper by Blum in 1967 [77,78] and has grown since. According to [79], which is an important review paper on the topic, "*curve-skeletons* are thinned 1D representations of 3D objects [...]" and are of relevance in fields such as visualization, computer graphics, computer-aided design, image processing, computational geometry and animation [79,80]. Curve skeletons contain important topology information about the original 3D object [79]. There exist a variety of approaches and implemented methodologies to generate curve skeletons from either two-dimensional (2D) or 3D objects. This subsection focuses on the so-called *skeletonization* of 3D objects by 1D representations. It is noted that skeletonization approaches that generate 2D representations of 3D objects are not presented in the following paragraphs.

Approaches for generation of 1D curve skeletons from 3D objects can be classified into four groups, namely, 1. thinning and boundary propagation, 2. distance field-based, 3. geometric, and 4. general-field functions [79]. Figure 2.4 depicts exemplary steps of generation of curve skeletons for each of the four groups previously mentioned groups.

1. The basic approach in thinning and boundary propagation is that voxels are removed from the boundary in each iteration until no more voxels can be removed without changing the topology of the object (see Figure 2.4 (a)). [79–81]
2. For distance field approaches various distance field functions may be used. A slice of a distance field of a chess piece is depicted in Figure 2.4 (b) as an example. In the figure blue corresponds to a small distance from the boundary whereas red corresponds to a greater distance from the boundary of the object. The ridges of such a distance field provide potential voxel candidates of the curve skeleton. Identification of voxel candidates and generation of the final curve skeleton may be done using a variety of different algorithms. [79]
3. Geometric approaches include medial axis-based and surface-based techniques such as those that use the Voronoi diagram, cores and m-reps, and shock scaffold.

Another important approach is the use of Reep graphs. Figure 2.4 (c) shows exemplarily how the approach can be used to generate a curve skeleton from a 2D object. [78,79,82]

4. In general-field functions, fields other that the distance transform are used for generation of curve skeletons. One such approach is to subject the boundary of an object to an electric charge. Then the electric potential is calculated and the repulsive force is calculated. Figure 2.4 (d) shows the repulsive field of an exemplary 2D object. The curve skeleton can then be generated by detection and connection of local extremes in the repulsive field. These approaches tend be less sensitive to boundary noise and tend to generate smooth curves while being computationally relatively expensive. [79,80,83,84]

For more general information on skeletonization and approaches to it, the reader is kindly referred to the book by Saga et al. [78], a state of the art report on skeletonization [80] and the previously mentioned review paper [79].

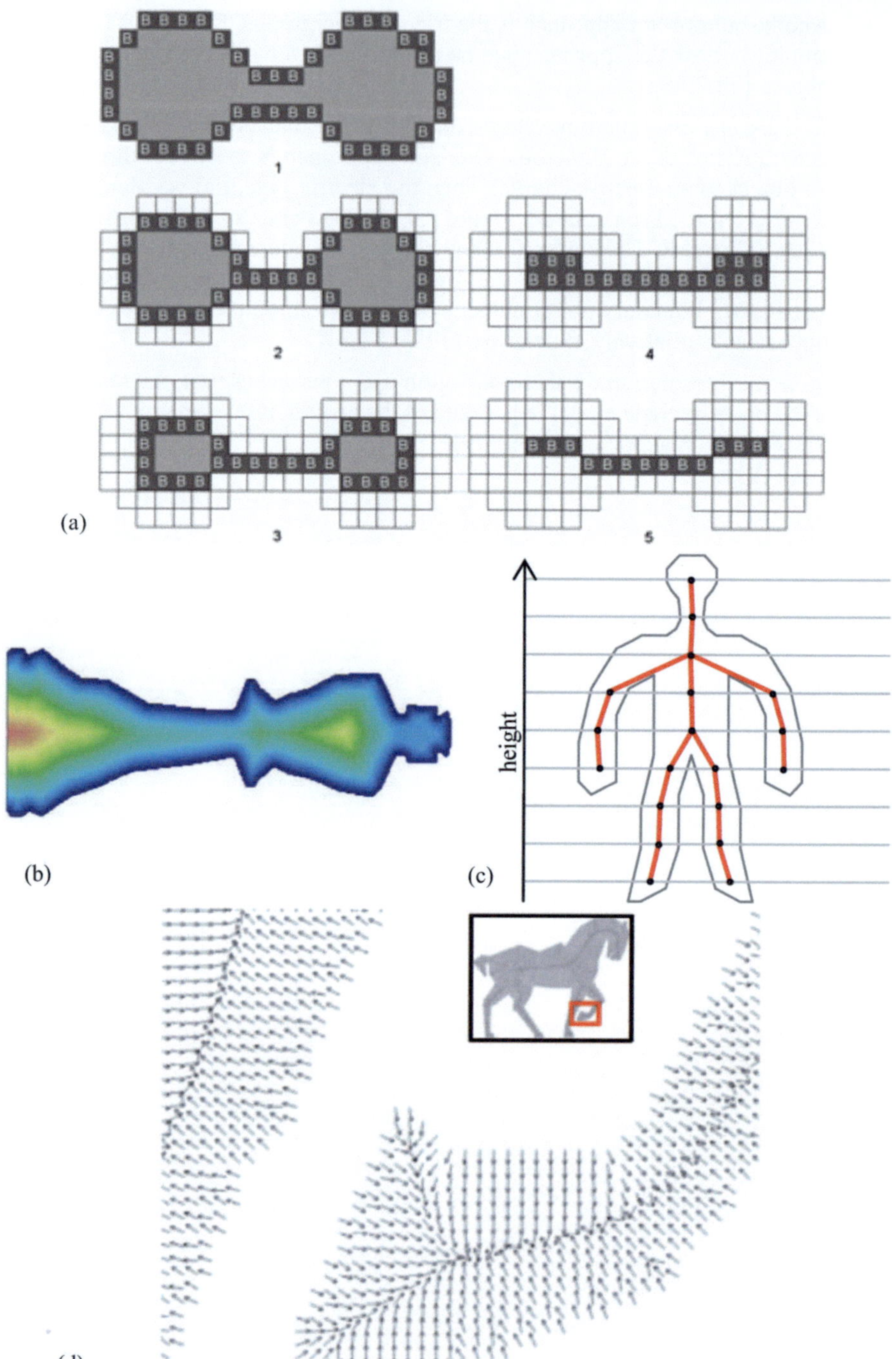

(a) (b) (c) (d)

Figure 2.4. (a): the thinning process on an example 2D shape. Boundary points are marked "B" at the beginning of each iteration and then removed if they are simple; (b): a color-coded slice of the distance field of a 3D chess piece shape; (c): an embedding of the Reeb graph into the original image space. Each node is taken to be the centroid of its corresponding contour; (d): the repulsive force field of a 2D shape. All four panels were reproduced with permission from [79], © IEEE, 2007.

Building on the research on curve skeletons and topology optimization, Nana et al. [85] and Cuillière et al. [76] focused on the generation of CAD models based on 3D topology optimization results. In their approach they generate a curve skeleton of the topology optimization result to generate a 3D CAD beam structure in a subsequent step. In the following, the approach presented in [76] is summarized.

The input for the methodology presented in [76] is a 3D topology optimization result based on the SIMP approach. First, Taubin smoothing iterations followed by Laplacian-type smoothing iterations are utilized to achieve an optimized smooth 3D shape (Figure 2.5 (a)). Then, a curve skeleton is generated based on the optimized smooth 3D shape using a mesh contraction Laplacian-based algorithm and a subsequent farthest-point sampling process (see Figure 2.5 (b)). In a next step, the mean local radius of the smoothed topology optimization radius can be evaluated along the curves of the curve skeleton (see Figure 2.5 (c)). In the next steps, the curve skeleton is modified to a version consisting of straight lines (see Figure 2.5 (d) for the so-called normalized skeleton) and is combined with the information on the mean local radius (see Figure 2.5 (e)). Next, a 3D model consisting of cylindrical beams is generated from the data (see Figure 2.5 (f)) to achieve a new component design. Finally, the generated 3D abstraction is evaluated by a mixed-dimensional FEA model with respect to its mechanical properties. In their work the authors apply their methodology successfully to three example cases. Furthermore, the authors of the study state that an optimization potential of their approach is given by the optimization of beam cross sections. For more information on the methodology the reader is kindly referred to the respective work [76].

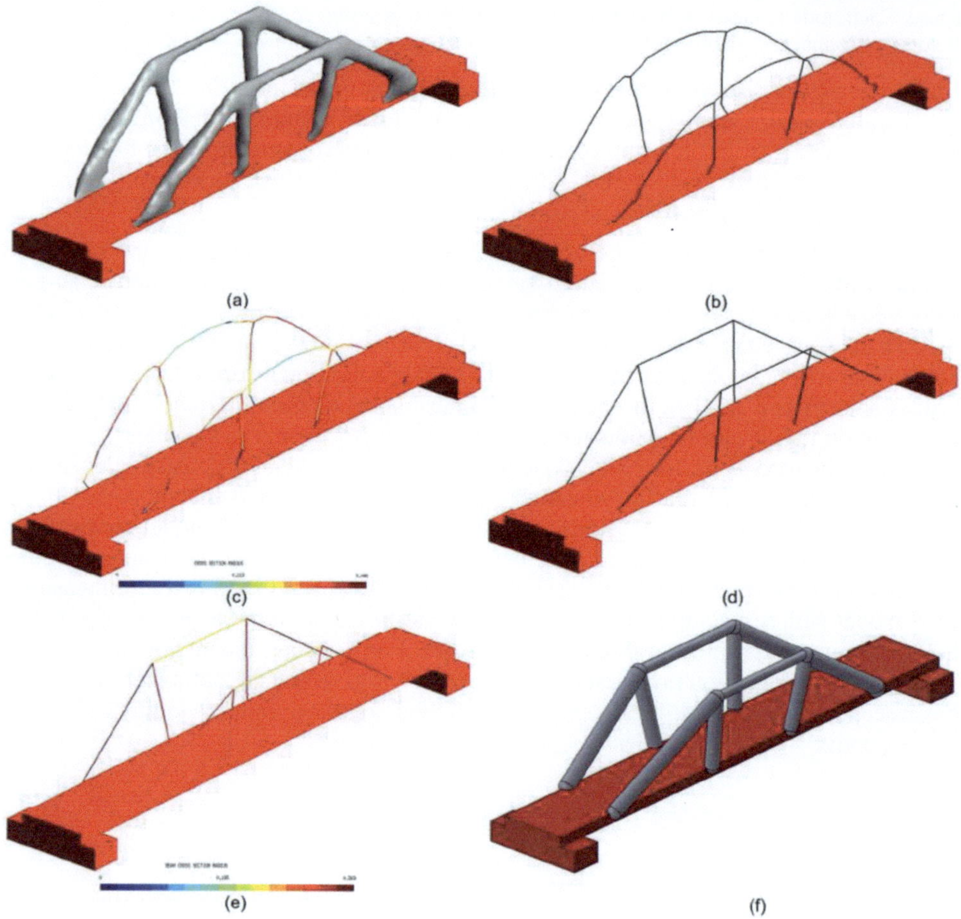

Figure 2.5. Skeletonization and abstraction of a 3D optimized solid shape. (a): 3D solid shape after smoothing; (b): curve skeleton derived; (c): distribution of cross section radius along skeleton branches; (d): skeleton after normalization; (e): distribution of beam cross section radii; (f): 3D CAD model of the optimized beam structure obtained at the end of the process. Reproduced with permission from [76], © U-turn Press LLC, 2017.

Similar approaches have been recently implemented in the commercial software Synera Platform [86]. In a YouTube video from late 2023 [87] Synera demonstrates how its software can be used to generate a solid beam network from an input topology optimization result. In the first step of the process, voxelization of the topology optimization result is performed. Next, a reduction to a curve skeleton is done. Based on the video contents it can be assumed that a thinning approach was implemented in the software. The generated curve network is then cleaned (similar to the normalization of the skeleton in [76]). Furthermore, certain curves are extended to the non-design spaces. In a next step, a solid beam network based on curve network is created. All beams of the network have the same radius. To achieve the final component design, unification of the solid beam network with the non-design spaces is carried-out and fillets are added to the design. It is noted that dimensioning optimization of the beams is not presented in the tutorial, but it is stated in the video that adding such an optimization is one option to extend the workflow. [87]

The paragraphs above provide a brief introduction into approaches and methods on how topology optimization results may be further developed into 3D component designs using skeletonization. In the introduction of this thesis, a study [10] was briefly presented, which combined topology optimization and biomimetic design principles to achieve a light-weight component design (see Figure 1.1). Motivated by this, the field of biomimetics will be presented in more detail in the following section.

2.3 Biomimetics

The term *biomimetics* was coined by Otto Schmitt who worked in the field since the 1950s and used it in the title of one of his works in 1969 [88,89]. The term *bionics* is closely related to biomimetics and was coined by Jack Steele in 1960 [89,90]. Although there is no clear consensus in academia, some experts consider biomimetics and bionics equivalent [91]. Consequently, this thesis uses the two terms interchangeably.

According to VDI 6220 biomimetics is defined as the

"interdisciplinary cooperation of biology and technology or other fields of innovation with the goal of solving practical problems through the functional morphological analysis of →biological systems, their →abstraction into →models, and the transfer into and application of these models to the solution" [92].

VDI 6224 [93] presents an integrated product development process for biomimetic optimization. This process begins with the assessment of component requirements and concludes with a CAD file that is ready for manufacture [93]. It is based on the Evolutionary Light Structure Engineering (EliSE) process [94].The proposed process combines topology optimization, biomimetics, parametric design optimization, validation by FEA, and AM [93] and provides a conceptual method for development of biomimetic structural components. It is noted that the methods are rather coarse-grained.

In the following, exemplary studies are presented in which biomimetics are applied together with topology optimization and AM. It is noted that in the field of AM component designs with curves and rounded edges often are referred to as biomimetic, bionic or organic [95]. Also, incorporation of lattice structures and use of topology optimization methods often results in a design being referred to as biomimetic [95]. According to [95] all these approaches are referred to as biomimetic in this thesis. Furthermore, such approaches or designs that use a direct input from nature or biology will be referred to as *true biomimicry* and *true biomimetics*.

Therefore, in this thesis topology optimization is considered a mathematical and biomimetic approach. However, depending on the specific approach and implementation of the topology optimization algorithms, topology optimization may be considered true biomimetic if e.g. the solver uses biomimetic principles or another direct biological input was used for generation of the topology optimization result.

2.3.1 Selected biomimetic AM components

In this subsection selected biomimetic AM components are introduced. Together Bugatti Automobiles S.A.S. and the Fraunhofer IAPT developed a novel AM brake caliper. Topology optimization, biomimetic design principles and a material change were combined to develop a new component that was 40 % lighter than the previous component design [96,97]. It is not clear from the references, whether the biomimetic design principles refer only to the application of topology optimization or refer to other potential biomimetic design principles.

In the introduction of this thesis an aerospace bracket was briefly introduced (see Figure 1.1) that was designed combining biomimetics, topology optimization and AM. Also another kind of biomimetic AM aerospace bracket was developed with the help of topology optimization that achieved a mass reduction of 13 % compared to the previous design of the component [58,98].

In general, biomimetic design that is based on biological input is still rare in DfAM [95]. This once more highlights the relevance of this thesis in which design methodologies for true biomimetic designs are presented. For more information on the overlap of the fields of biomimetics, AM, and topology optimization and additional examples of applications the reader is kindly referred to the two review articles [11] and [95].

2.3.2 Biomimetic beams

In this subsection, a brief introduction into the topic of biomimetic beams is given, as they are of relevance for this thesis. It is noted that this subsection is partly based on one of the central papers [25] of this thesis.

Considering forces act on the ends of a beam only, a beam may be subject to normal forces, shear forces, torsion (moment around the beam's main axis) as well as bending moments. Tensile stiffness and shear stiffness are geometrically dependent on the cross-sectional area and, therefore, have low relevance for geometric optimization of beams with constant cross section. Considering the approach of this thesis, torsion does not have to be expected to be of high relevance. However, relevant factors in beam design for this thesis are bending stiffness and buckling resistance. The bending stiffness is given by $E \cdot I$ with the Young's modulus of the material of the beam E and its area moment of inertia of the cross section about the axis of bending I [99]. The critical force F_{crit} of a slender beam under buckling that is pinned at both ends is given by

$$F_{\text{crit}} = \pi^2 \frac{E \cdot I}{l^2} \tag{2.3}$$

with E, I, and the length of the beam l [99,100]. Therefore, both buckling resistance and bending stiffness are strongly dependent on I. Apart from overall buckling and bending stiffness, another aspect that needs to be considered is local buckling of e.g. thin walled structures that might occur [101–103].

Many biological structures may be considered in the development of biomimetic beams such as bamboo, grass stems, horsetail, porcupine quill, hedgehog spine, bird nests, cocoons, trees, turtle shells, bone, cacti, cattail, palms, lemon, and the banana plant pseudo-stem [102–107]. With regard to bending stiffness and buckling resistance that are of relevance for the thesis at hand, grass-stalks, horsetail, bamboo, and the banana plant pseudo-stem were found to be of great interest.

In a book on biological design [107] the cut surface of the pseudo-stem of a banana plant (see Figure 4.10 (b) for a photo) is described as layered leaves. The authors describe the lightweight design characteristic of the leaves as strongly reinforced outer skins (epidermises) [107]. The epidermises are connected by struts resulting in a sandwich construction [107]. Grass-stalks and horsetail carry a ring-like biological supporting frame at their outside [104]. This stiffening structure contribute to a great area moment of inertia and contributes to their great bending stiffness [104]. Bamboo is a biological structure that is known for its considerable buckling resistance, strength, and flexibility [102,104]. A study [102] developed a biomimetic beam design based on bamboo and assessed the structure with regard to elastic buckling.

3. Research questions, methodology, and materials

The current gap in research for biomimetic design for AM was presented in the introduction and is given by a lack of design tools for the development of detailed biomimetic component designs that can be produced via AM. As it was presented in the previous chapter, topology optimization approximates optima for mechanical problem formulations. However, a limitation is given by the mesh-dependency of results and, therefore, the level of refinement of the finite element meshes. Usually better solutions can be found in density-based topology optimization if mesh refinement is high. Therefore, a limitation is imposed by the available computational resources. Increased geometric complexity due to the integration of biomimetic beams might improve a component's lightweight characteristics.

3.1 Research questions

Based on these considerations, the following primary and secondary research questions were formulated as a foundation for this thesis:

Q1: Primary research question

How can a biomimetic design methodology be developed to redesign mechanical components generated from density-based topology optimization by substituting cylindrical beams with biomimetic beams, while considering manufacturability by PBF-LB/M?

Q2: Secondary research question

How do biomimetic designs generated with such a methodology compare to the original topology optimization designs, in terms of lightweight characteristics?

Previous studies have shown that incorporation of biomimetic beams based on e.g. bamboo can improve the lightweight character of component designs [10]. Based on this example this thesis focuses on the development and assessment of methodologies for biomimetic design of structural components based on topology optimization. The more general approach for a design methodology developed within this thesis is presented in chapter 4.

3.2 Methodology

In this section, the general approach to investigate the two research hypotheses is presented. The primary research question (Q1) is investigated by the development of respective design methodologies. The development is based on the V-model that is of importance in the product development of mechatronic and cyber-physical systems according to VDI/VDE 2206 [108].

T. Röver, *Methodology for Biomimetic Design of Additively Manufactured Structural Components*, Light Engineering für die Praxis, https://doi.org/10.1007/978-3-032-27405-2_3

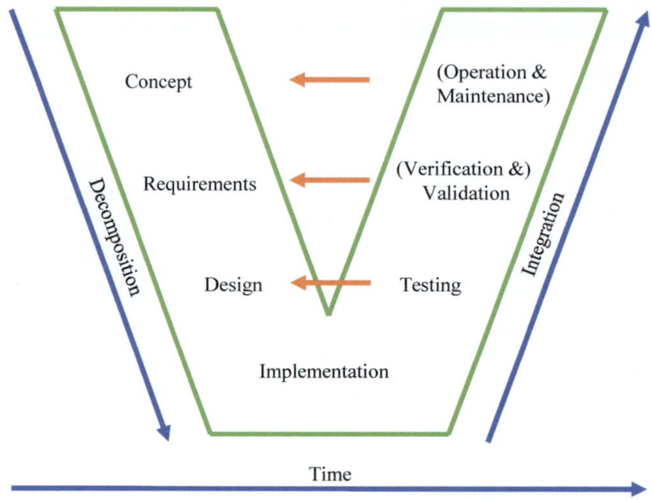

Figure 3.1. Schematic of the V-model based on [109]. Steps in brackets were not performed within this thesis.

Figure 3.1 shows a schematic of the V-model. In the three steps on the left side of the schematic, the concept, requirements, and design correspond to the decomposition part of the model. In this part the product is broken down from a high conceptual level to detailed tasks. In the most detailed part at the bottom of the V stands implementation. The three steps on the right, namely testing, verification & validation, and operation & maintenance correspond to the integration part of the model. [109]

According to the first step of the V-model a concept for a design methodology was developed. Next, requirements for the methodology were defined. The steps of design, implementation, and testing were performed for various submethodologies. These steps mainly correspond to the investigation of Q1.

In a next step, validation of the methodologies was performed by applying them to three exemplary cases. Assessment of the results allowed for the investigation of Q2. It is noted that the steps of verification and operation & maintenance were not considered within this thesis as they were not necessary for investigation of the two research questions.

3.3 Materials

In this section, the software, materials, and machines used in this work are stated. COMSOL Multiphysics 5.4 (COMSOL 5.4) [110] as well as COMSOL Multiphysics 6.0 (COMSOL 6.0) [111] were used. Java JRE 8 and Java JRE 11 were used in the so-called *application builder* of COMSOL 5.4 and COMSOL 6.0, respectively. As an addition, the JAMA package for Java [112] was used.

Furthermore, Autodesk Inventor 2024 (Inventor 2024) [113], Autodesk Netfabb Ultimate slicer software (version 2023.1) (Netfabb) [114], and nTop 4.22.2 (nTop) [115] were utilized. For AM the PBF-LB/M system MPrint from One Click Metal GmbH [116] was used.

Table 3.1 summarizes the relevant material properties of the aluminum alloy AlSi10Mg that was used as a material of choice in wide parts of this thesis. AlSi10Mg was comprehensively studied in research with regard to its use in PBF-LB/M [117,118] and shows beneficial properties with regard to its thermal conductivity, strength-to-weight ratio, weldability, shrinkage, and mechanical properties [117]. Al-Si-Mg alloys are of interest for industries and applications such as automotive, aerospace, motor racing, and heat exchangers [118]. This contributed to the decision to focus on AlSi10Mg in this thesis.

Anisotropy of material properties in metals manufactured by PBF-LB/M commonly have to be considered. In this thesis, AlSi10Mg manufactured in PBF-LB/M with a heat-treatment of 2 h at 300 °C was considered. Following the argument in [25], the yield strength and Young's moduli in the different directions in space are sufficiently close for heat-treated AlSi10Mg such that it is considered isotropic in this thesis. Furthermore, the maximum acceptable stress value is defined as an acceptance criterion denoted as $\sigma_{\mathrm{acc,AlSi10Mg}}$, which is 80 % of the yield strength of the material, at 112.8 MPa.

Table 3.1. Material properties of heat-treated AlSi10Mg manufactured by PBF-LB/M. Modified and reproduced from [25] (licensed under CC BY 4.0).

Material property	Symbol	Value
Yield strength	$\sigma_{\mathrm{y,AlSi10Mg}}$	141 MPa [119]
Young's modulus	E_{AlSi10Mg}	57 GPa [119]
Poisson's ratio	ν_{AlSi10Mg}	0.34 [120]
Density	ρ_{AlSi10Mg}	2.67 g/cm^3 [119]

4. Biomimetic design methodology

In this chapter, the biomimetic design methodology that was developed in this thesis is presented in detail. It is noted that large parts of the contents were already published [23–26]. The four articles can be found in the appendix of this thesis. This chapter is largely based on these publications, the respective data sets [121–123], and four student theses [61,124–126] and summarizes them. For more details on the works the reader is kindly referred to the respective publications, datasets, and student theses. As stated in chapter 3, the procedure was based on the V-model.

4.1 Concept solution

In a first step, a conceptual solution for the design methodology was developed. Based on the example of the FCRC bracket [10] (see Figure 1.1), literature [93], a previous work [61], and the research hypotheses, a concept solution for the development of a biomimetic design methodology was developed. A concept on the input and output of the design methodology is shown in Figure 4.1. The input is given by the result of a density-based topology optimization. The output of the biomimetic design methodology is a biomimetic component design. The component design is based on the topology optimization result, contains biomimetic features, and is manufacturable by PBF-LB/M.

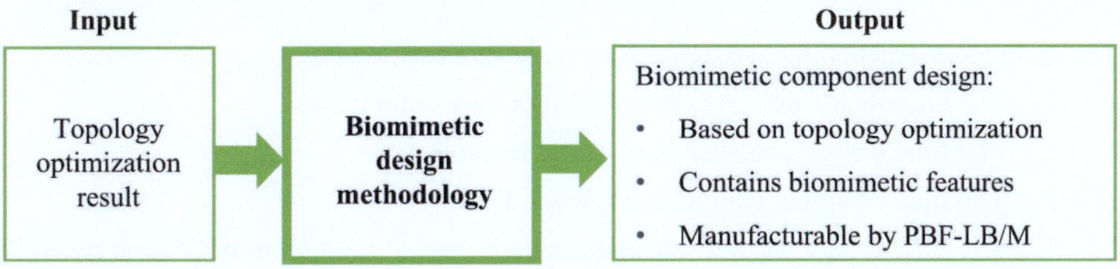

Figure 4.1. Concept on input and output of the biomimetic design methodology. Partly based on [61].

In Figure 4.2, a schematic of the conceptual approach on an exemplary two-dimensional (2D) model is depicted. On the left side the result of a density-based topology optimization for a common mechanical problem formulation, namely the L-shape, can be found. The schematic output is given by a component design in which the beam-like segments of the topology optimization result are substituted by biomimetic beams and the nodes of the structure are adapted in the newly developed design. As part of this thesis, the concept was first introduced in [23].

T. Röver, *Methodology for Biomimetic Design of Additively Manufactured Structural Components*, Light Engineering für die Praxis, https://doi.org/10.1007/978-3-032-27405-2_4

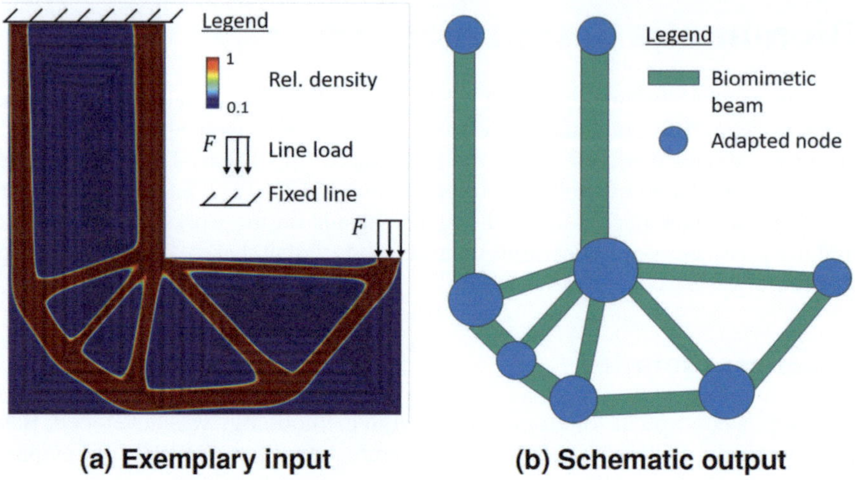

Figure 4.2. Schematic of the proposed approach to convert a result from density-based topology optimization into a component composed of biomimetic beams and adapted nodes. Reproduced from [23] (licensed under CC BY 4.0).

4.2 Requirements

The main requirements in development of the design methodology were defined as:

1. The biomimetic design is based on topology optimization
2. The final design contains biomimetic features
3. The final design is manufacturable via PBF-LB/M
4. The final design has a lower mass than the input topology optimization result
5. The final design can support the acting loads
6. The methodology allows for a good reproducibility of the design process
7. The methodology allows for a "trackable and easily-documentable component development process" [23]
8. The methodology allows for "[t]he possibility of automating the design process" [23]

It is noted that [23] does not contain this exact list of requirements. However, all listed requirements were discussed directly or indirectly in [23] and were considered in the development of the methodologies presented in this thesis.

4.3 A five-step methodology

In this section, key considerations regarding the evaluation of internal forces and moments are presented. Furthermore, the conceptual five-step methodology that was developed in this thesis is presented.

4.3.1 Considerations regarding evaluation of internal forces and moments

Incorporation of biomimetic beams is one of the main aspects of the approach presented in this thesis. To develop and dimension biomimetic beams that are suitable for incorporation into the component, the length of each beam and the loads acting on each beam must be known. Figure 4.3 shows an exemplary topology optimization result and an abstraction of the topology optimization result by cylindrical beams and spherical nodes. In FEA, evaluation of internal forces and moments can be performed on so-called cut surfaces. The evaluation of forces and moments at two cut surfaces in each beam is sufficient to allow for the dimensioning of the biomimetic beams if the beams are straight and have a continuous cross section. More details on this aspect are presented in section 4.6 .

One possible approach might have been to discretize the TO volume model (see Figure 4.3 top left) and evaluate internal forces in the beam-like segments directly. However, this approach was found to be disadvantageous. The irregular shape of the beam-like segments was expected to be problematic in evaluation of the internal forces. Furthermore, approximations for the main axis and the start and end points of each beam-like segment would be necessary. Combining such approximated geometrical data with the irregular-shaped TO volume model was expected to give rather imprecise values for the internal forces and moments.

Therefore, it was found to be beneficial to develop approximations for the main axes and the start and end points of each beam-like segment and generate a 3D auxiliary model from the data (see Figure 4.3 top right for an example). Evaluation of internal forces in geometrically well-defined beams of such an auxiliary model was expected to be more precise compared to the previously presented approach.

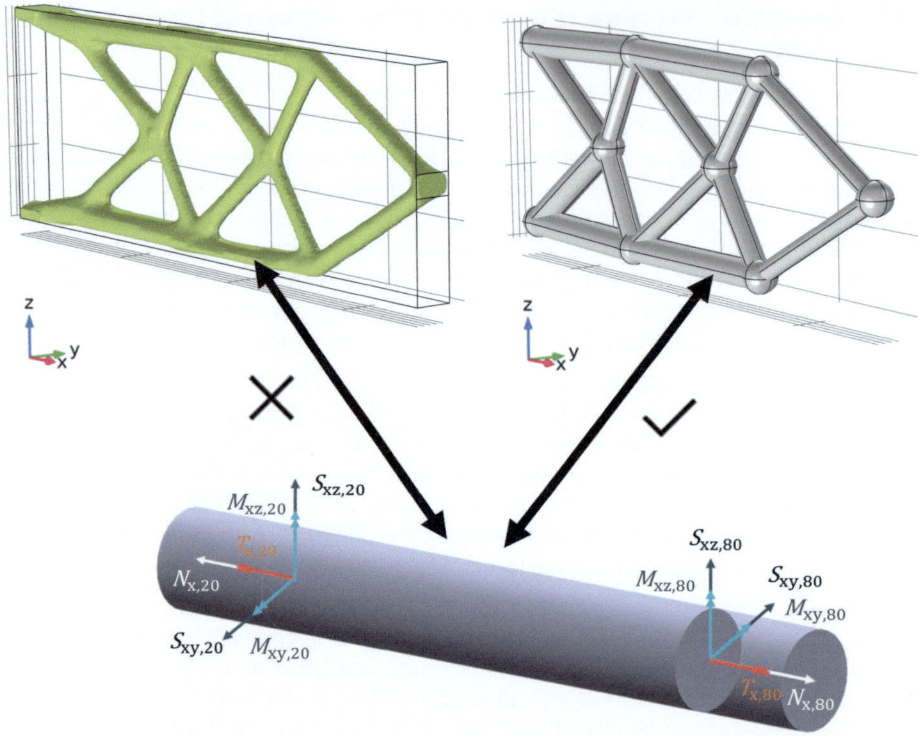

Figure 4.3. Schematic indicating more error-prone identification of beam geometry and relevant cross sections in topology optimization result (top left) and the easier identification of beam geometry and relevant cross sections in an auxiliary model (top right). Beam with internal forces and moments at two cross sections (bottom). Modified and reproduced from [23] (licensed under CC BY 4.0) and [24] (licensed under CC BY 4.0).

4.3.2 The conceptual five-step methodology

Based on the previously presented concept solution, requirements, and considerations, a more detailed five-step methodology was developed and presented in [23]. This methodology represents a first version of the biomimetic design methodology that was developed throughout this thesis. For more information on this version of the concept solution the reader is kindly referred to [23]. Throughout this thesis, the five-step methodology was refined and improved. The latest version of the biomimetic design methodology was presented in [26]. Its schematic is depicted in Figure 4.4.

The following paragraphs introduce the conceptual methodology and are strongly based on [23] and [26]. The input for the methodology is given by a 3D topology optimization result and its boundary conditions.

In a first step, a skeletonization is performed to achieve an abstraction of the topology optimization result by points and lines. In a next step, an auxiliary 3D model of the component is developed by cylindrical beams and spherical nodes. In step two, an FEA of the auxiliary model based on the mechanical boundary conditions of the topology optimization is performed. In post-processing of the FEA, internal forces and moments in the beams of the abstraction are evaluated and post-processing of the data is performed.

In a third step, a submethodology for selection and dimensioning of biomimetic beams is used. For each member of the auxiliary model, a biomimetic lightweight beam is selected and a corresponding beam design is generated. This submethodology contains parameter optimization, beam selection and buckling analysis. In step four, nodes that connect the biomimetic beams are generated to obtain complete biomimetic component designs. A general shape of each node is generated and a lattice structure is introduced into the nodes. Furthermore, powder removability based on PBF-LB/M is ensured during the design process.

In the final step, an FEA is run for validation of the design. If necessary, feedback loops to earlier steps of the methodology are used if the FEA reveals the existence of critical regions of the design in terms of maximum occurring stresses. In case the design supports the loads, the mass of the component is evaluated in the 3D model to assess its lightweight character.

Detailed solutions to each of the steps are presented in the following sections of this thesis. They contain information regarding the steps of design, implementation and validation according to the V-model (see section 3.2).

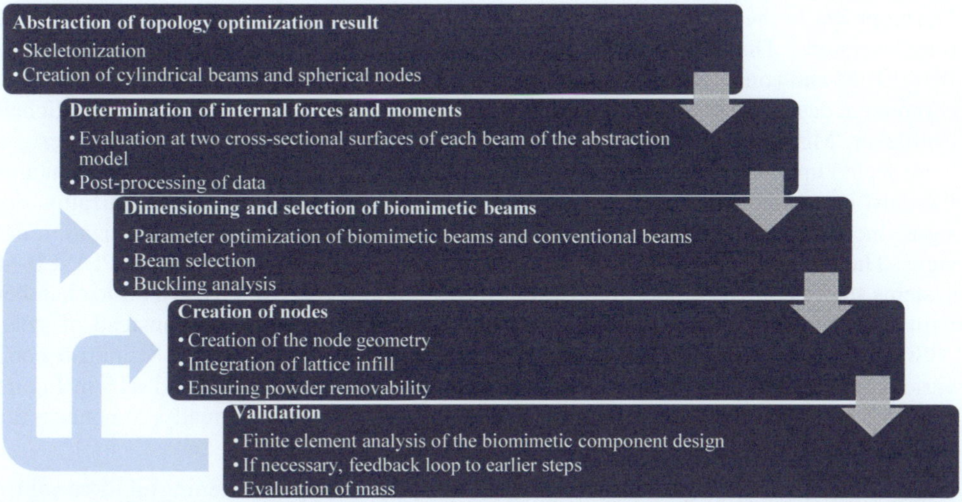

Figure 4.4. Flowchart of methodology for development of additively manufactured structural components through biomimetics based on topology optimization results. Modified and reproduced from [123] (licensed under CC BY 4.0).

4.4 Input topology optimizations

In this section, the generation of input topology optimization results is presented. This section is largely based on [24] and [26]. In a first step, three mechanical problem formulations that are commonly used in the field of topology optimization, namely Michell cantilever, L-shape, and Messerschmitt-Bölkow-Blohm (MBB) beam were identified. The problem formulations were based on [127] and [128]. Two versions of topology optimization results for each problem formulation were generated. The initial and the final versions are presented in more detail in [24] and [26], respectively.

4.4.1 Initial input designs

For the generation of the initial versions of the topology optimization results, design spaces were generated and the respective mechanical boundary conditions were introduced based on previous studies [127,128]. Structural steel (e.g. ASTM A36) with a Young's modulus of 200 GPa, a Poisson's ratio of 0.30, and a density of 7850 kg/m^3 was considered. The selection of structural steel and values of Young's modulus and Poisson's ratio were based on [127]. For the topology optimizations, hexahedral finite elements were used. The software COMSOL 5.4 and the SIMP method with a Helmholtz-filter as implemented in the software were used. Furthermore, an additional boundary condition was implemented to prevent material to be distributed close to the outer surfaces of the design spaces. Preliminary studies within this thesis showed that this approach was beneficial for an easy export of the topology optimization results in the software. Finally, the three topology optimizations were performed and three volume models could be generated and exported as STL files. The files can be found in the respective data set [121].

4.4.2 Final input designs

Development of the final versions of the topology optimization designs was based on the initial versions. The designs were scaled up. Due to the manufacturing restrictions of PBF-LB/M and considering the design approach of this thesis, it was assumed that larger component designs show a higher potential for improvement. The STL files of the Michell cantilever, MBB beam, and L-shape were imported into COMSOL 6.0 and the designs were scaled up geometrically by factors of 4, 4, and 1.5, respectively. Table 4.1 contains the most important information regarding the upscaling of the topology optimization designs, including the approximate dimensions of the upscaled topology optimization designs. The upscaled component dimensions are in line with medium to large PBF-LB/M systems such as the SLM 800 from Nikon SLM Solutions AG, which has a build chamber with dimensions of 500 mm × 280 mm × 850 mm [129]. Additionally, instead of steel, AlSi10Mg was assigned to the three models based on its relevance in AM industry and various applications (see section 3.3). Additionally, two beams (#15 and #16 in Figure 4.18 (b)) were added to the abstraction of the Michell cantilever manually, to connect the abstraction with its mechanical supports.

In the next step, upscaling of the acting forces was done to acquire meaningful force values for the modified topology optimization designs. Parameter optimizations were developed and performed for all three updated topology optimization designs in COMSOL 6.0. An optimization constraint was given by the maximum allowable stress of $\sigma_{acc,AlSi10Mg}$ (112.8 MPa). It is noted that stress evaluation in the optimizations was omitted in regions close to the mechanical boundary conditions to generate physically meaningful results. The force values have been upscaled in the optimization to 8237.5 N, 16066 N, and 2981 N for Michell cantilever, MBB beam, and L-shape, respectively. The upscaled topology optimization designs and their upscaled mechanical boundary conditions represent the final topology optimization inputs that were used in this thesis for the application of the biomimetic design methodology.

Table 4.1. Geometric scaling factors, approximate dimensions of upscaled topology optimization designs, and identified loads of upscaled topology optimization designs. Modified and reproduced from [123] (licensed under CC BY 4.0).

Model	Geometric scaling factor	Approximate dimensions of upscaled design [mm^3]	Load [N]
Michell cantilever	4	640 × 240 × 40	8237.5
MBB beam	4	800 × 160 × 40	16044
L-shape	1.5	300 × 300 × 15	2981

4.5 Abstraction of topology optimization result

Section 4.3 discussed why an abstraction of the input topology optimization should be done to efficiently evaluate the internal forces and moments in each of the beam-like sections of a topology optimization design. In this section, methodologies for the abstraction of topology optimization results are presented. The abstraction is given by a solid volume model consisting of spherical nodes and cylindrical beams. It is noted that this section is largely based on three central papers of this thesis [23,24,26] and summarizes relevant contents of the works.

Methodologies and approaches to create solid volume models from topology optimization results were briefly introduced in subsection 2.2.2. Skeletonization was found to be a useful tool to achieve such 3D abstractions. This is why skeletonization was used in this thesis for abstraction. Therefore, in a first step the most suitable skeletonization approach had to be chosen. According to literature [79] general-field functions tend be less sensitive to boundary noise and tend to generate smooth curves. Furthermore, when reviewing results from the respective article [79], the implemented general-field approach seemed to effectively identify the most important nodes of the curve skeletons. This is why general-field functions and more precisely the potential field approach were chosen in this thesis, despite being relatively expensive with regard to computational costs [79].

Figure 4.5 shows the steps of the abstraction methodology that is proposed in this thesis. More detailed information on the steps is given in the following subsections. It is noted that the developed abstraction methodology is inspired by and partly based on [76,83–85].

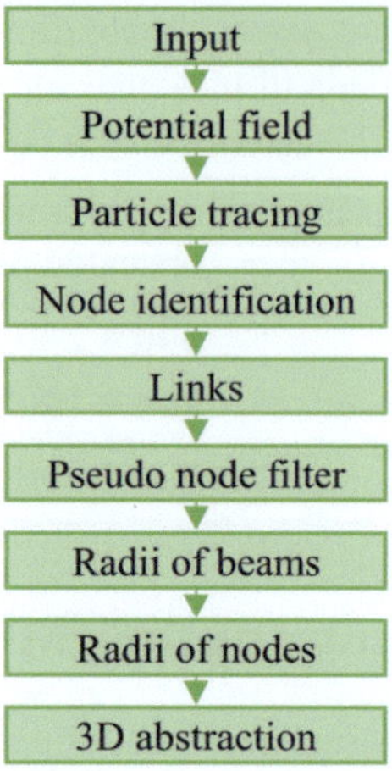

Figure 4.5. Overview of the steps of the abstraction methodology. Reproduced from [24] (licensed under CC BY 4.0).

4.5.1 Input and potential field

According to the overall biomimetic design methodology, the input for the abstraction methodology is given by a 3D volume model of a topology optimization result and its mechanical boundary conditions. An exemplary 3D volume model (Michell cantilever) can be found in Figure 4.6 a). In the next step, following the potential field approach, a negative electrical potential is applied to the surface of the 3D volume model and a positive space charge density is applied to its volume. FEA is performed to calculate the electric field (see Figure 4.6 b)) with low potential near the surface and increasing potential toward the inside of the volume model. The arrows in the figure show the gradients that point from the inside of the volume model to the outside.

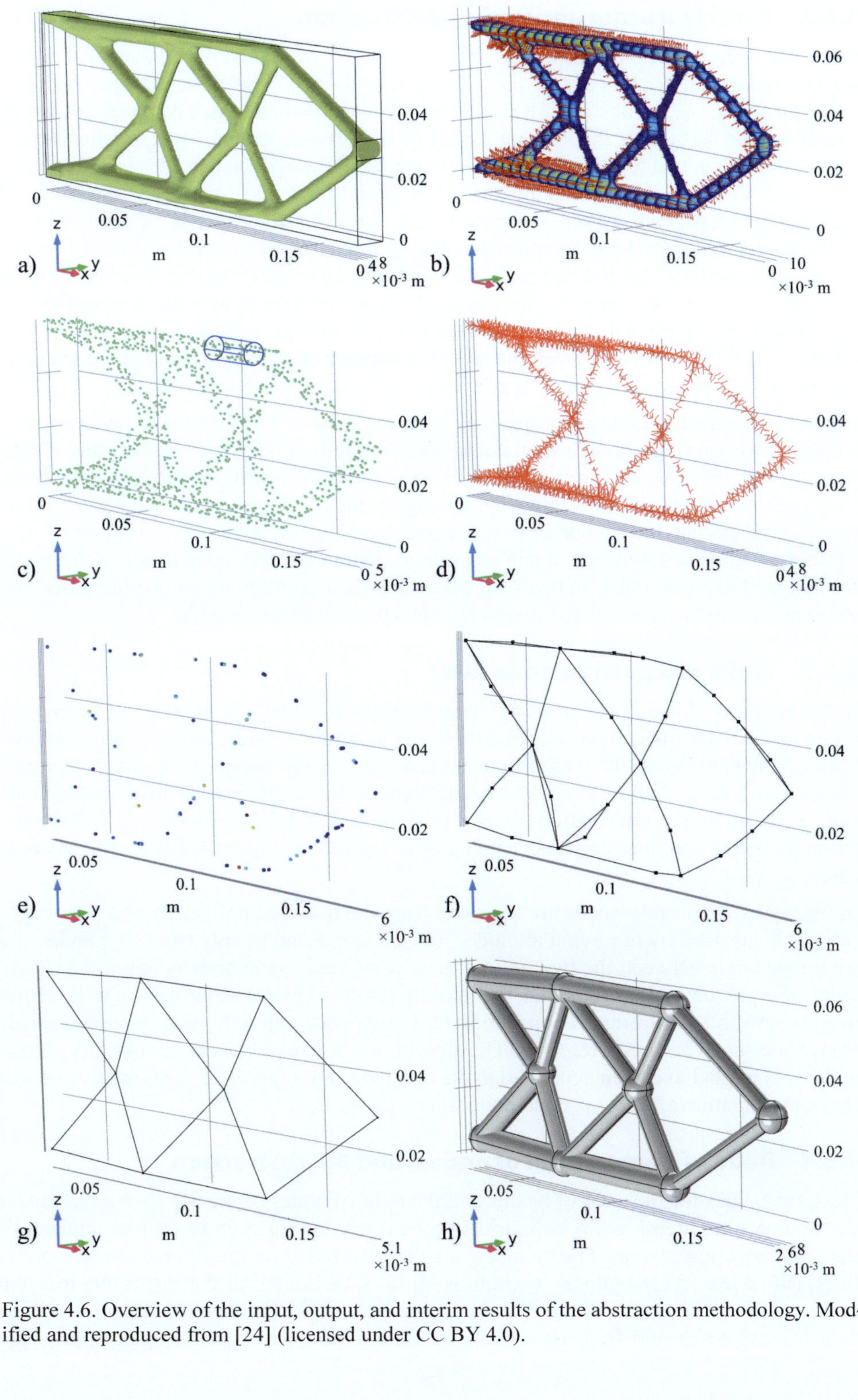

Figure 4.6. Overview of the input, output, and interim results of the abstraction methodology. Modified and reproduced from [24] (licensed under CC BY 4.0).

4.5.2 Particle tracing and node identification

In the next step, a particle tracing study is performed to generate relevant data, that allows for the identification of nodes of the volume model as well as links between nodes in subsequent steps of the abstraction methodology. Negatively charged particles are spread evenly on the surface of the volume model of the time-depended FEA. Furthermore, a friction force is added to the simulation such that oscillation of the particles is reduced and they grind to a halt within the simulation. The initial position of the particles (green) of the example model can be seen in Figure 4.6 c). Figure 4.6 d) indicates trajectories of the particles and in Figure 4.6 e) the position of the particles at the end of the time-dependent study can be seen. It is noted that multiple particles may be so close that they appear to be a single particle in the figure. Numerous particles accumulate in various areas of the volume model. The particles can be classified into two groups. Firstly, those particles that grind to a halt in a node of the volume model, and those that grind to a halt along the main axis of one of the beam-like regions of the volume model which link nodes.

For node identification, the positions of all particles at the end of the time-dependent study are analyzed. Starting with the final coordinates of the first particle, the algorithm identifies all other particles within a defined distance and allocates the particles to this node. Running through all particles this way, the algorithm checks whether or not the particle was already assigned to a node or not. If a node was already assigned, it is skipped. If it was not assigned to a node yet, again all particles within the defined distance of this node, are assigned to a new node. In this way, nodes and their coordinates can be identified and information can be gathered about which node a particle is running into.

4.5.3 Links and pseudo node filter

In the next step, links between nodes must be identified. This is done by analyzing the initial coordinates of each particle and their assignment to a node. All possible combinations of any particles initial coordinates are analyzed. If any two particles are sufficiently close to each other and the run into two different nodes, a link between two nodes is assumed. Based on the information up until this point within the methodology, a representation by points and lines can be generated as is depicted in Figure 4.6 f) for the example case.

In the next step, pseudo nodes are removed from the data by application of a filter algorithm. This is done by removing all nodes that are connected to only two other nodes and where the angle between the two linking lines connected to the node is within 153 ° and 190 °. If a pseudo node is removed, its links are removed by the algorithm as well. Moreover, a new link is introduced connecting those two nodes directly, that the pseudo node was connected to before its removal. The algorithm is run two times to remove any pseudo nodes that might have been generated in the first iteration. Figure 4.6 g) shows a point and line representation after the pseudo node filter was applied.

4.5.4 Radii of beams, radii of nodes, and 3D abstraction

Next, the values for the radii of beams and the radii of nodes of the 3D abstraction under development are identified. A cylindrical probe is created for each beam-like segment of the topology optimization. The cylinders are oriented based on the filtered abstraction by lines and nodes. Every cylinder's length is 60 % of the length of the respective line and centered on it. The radius of the cylinder probe is set to a value based on the overall dimensions of the design (e.g. 10 mm). A cylindrical probe of one exemplary beam-like segment can be seen in Figure 4.6 c). To determine the beam's thickness, the coordinates of all particles within the cylindrical probe at the beginning of the time dependent particle tracing study are considered. A regression line based on all particle coordinates within the probe is calculated. Next, the distance between the initial particle positions within the probe and the regression line are calculated. The average of all distances is set as the radius of the respective beam for the abstraction.

The radii of the spherical nodes of the abstraction are set at 110 % of the greatest radius of any of the beam radii as was identified in the previous step of the abstraction methodology. In the final step of the abstraction methodology, a 3D abstraction consisting of cylindrical beams and spherical nodes is generated. Figure 4.6 h) shows the abstraction of the Michell cantilever. For more details on the abstraction methodology, the reader is primarily referred to [24,121] and secondarily referred to [23,26].

4.6 Evaluation of internal forces and moments

This section focuses on the evaluation of the internal forces and moments based on the 3D abstraction of a topology optimization by cylinders and spheres. In this step, the abstraction is meshed and an FEA is performed. It is noted that this section is largely based on [23,24,26] and summarizes relevant contents of the works as part of this thesis.

4.6.1 Fundamental considerations

At the bottom of Figure 4.3, a beam and its cut surfaces at 20 % of its length and at 80 % of its length are depicted. It is assumed that normal forces, shear forces, and torsion (moment around the beam's main axis) are constant along the length of each beam as forces and moments only act at the end of the beams. The bending moments within the beam are considered to consist of two parts. The first part is constant along the length of a beam and is due to bending moments acting on its ends. The second part is linear along the length of a beam and is due to shear forces at the beams ends.

4.6.2 Location of cut surfaces

Stress distributions might be complex near any node, which is why cut surfaces at 20 % and 80 % of the length of the beam were defined as the default locations of the two cut surfaces of any beam to be analyzed. However, it might be that beams that intersect at an acute angle result in cut surfaces that evaluate stresses in a neighboring beam. If that is the case, the locations of the cut surfaces are to be changed to a more suitable position. In Figure 4.7 beams with problematic locations of cut surfaces are depicted (see Figure 4.7 (a)) as well as an improved situation with more suitable locations of the cut surfaces (see Figure 4.7 (b)).

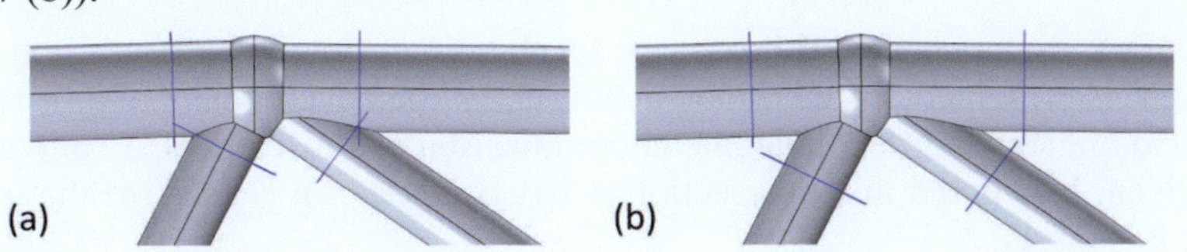

Figure 4.7. Exemplary critical node with cut surfaces for evaluation of internal forces and moments in abstractions. (a): planes for evaluation at 20 % and 80 % of lengths of beams that intersect with planes of neighboring beams; (b): alternative and preferred locations of cut surfaces for evaluation. Reproduced from [123] (licensed under CC BY 4.0).

4.6.3 Evaluation of forces and moments based on FEA

For the evaluation of the forces and moments in the cut surfaces, an FEA of the abstraction model consisting of cylindrical beams and spherical nodes has to be subjected to the original mechanical boundary conditions. Meshing of the 3D model is done. Then an FEA can be performed in COMSOL 6.0. Next, the cut surfaces in all beams may be evaluated using section forces as implemented in COMSOL 6.0 [130]. Figure 4.8 shows exemplary cut surfaces within an FEA model. For each cut surface the axial force N, the shear force along the first local axis T_1, the shear force along the second local axis T_2, the twisting moment M_t, the bending moment around the first local axis M_1, and the bending moment around the second local axis M_2 can be evaluated.

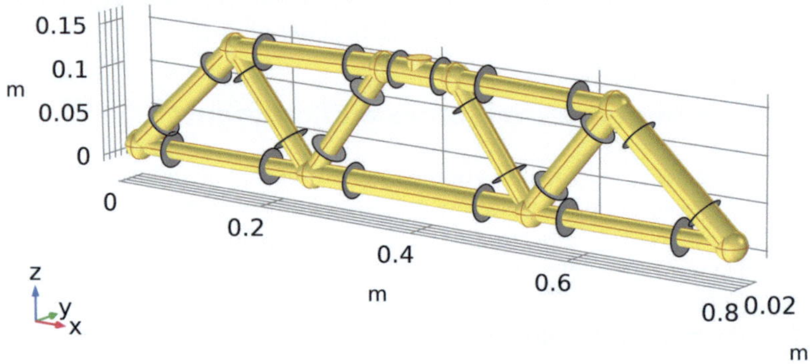

Figure 4.8. Screenshot of an exemplary FEA model of a 3D abstraction including cut surfaces (gray) in which internal forces and moments can be evaluated. © Röver, 2026.

In the next step, post-processing of the data has to be done to acquire the final values needed for dimensioning of biomimetic beams. The axial force N, the shear force along the first local axis T_1, the shear force along the second local axis T_2 and the twisting moment M_t were assumed to be constant along the length of the beam. Therefore, the average of each two values of a kind is taken as the final value for the next step. Doing so accounts for small deviations that can be expected due to the numerical approach and reduces their impact.

The bending moment around the first local axis M_1 and the bending moment around the second local axis M_2 were assumed to change linearly along the length of a beam. Based on the location of the cut surfaces along the length of a beam, values for M_1 and M_2 are extrapolated toward its ends.

4.7 Dimensioning and selection of biomimetic beams

In this section, the concept and procedure for the dimensioning and selection of biomimetic beams is presented. Furthermore, details about the development of a corresponding solution and implementation of parameter optimization models, and FEA validation methods are given. It is noted that this section is largely based on [23,25,26] that are central papers of this thesis.

4.7.1 Concept

In Figure 4.9 a schematic of the concept solution for the dimensioning and selection of biomimetic beams that was developed for this thesis is depicted. The input for the submethodology is given by relevant data of one of the beams of an abstraction model. This data includes the length of the beam according to the abstraction model as well as the evaluated (internal) forces and (internal) moments of the respective beam. In the next step, a dimensioning tool is used. Parameter optimization simulations for four different beam types are performed. The parametric beam designs consider the manufacturing constraints of the PBF-LB/M process. Furthermore, maximum occurring von Mises stresses and buckling criteria are considered to ensure that the developed beams can support the acting loads. It is noted that four parametric beams were considered in this thesis even though the methodology could be performed with less or more parametric beam types analogously. More information on the development of the four parametric beam designs is given in the following subsection. The output of the submethodology is given by a single beam design that supports the acting loads and whose length matches the length of the beam in the abstraction model.

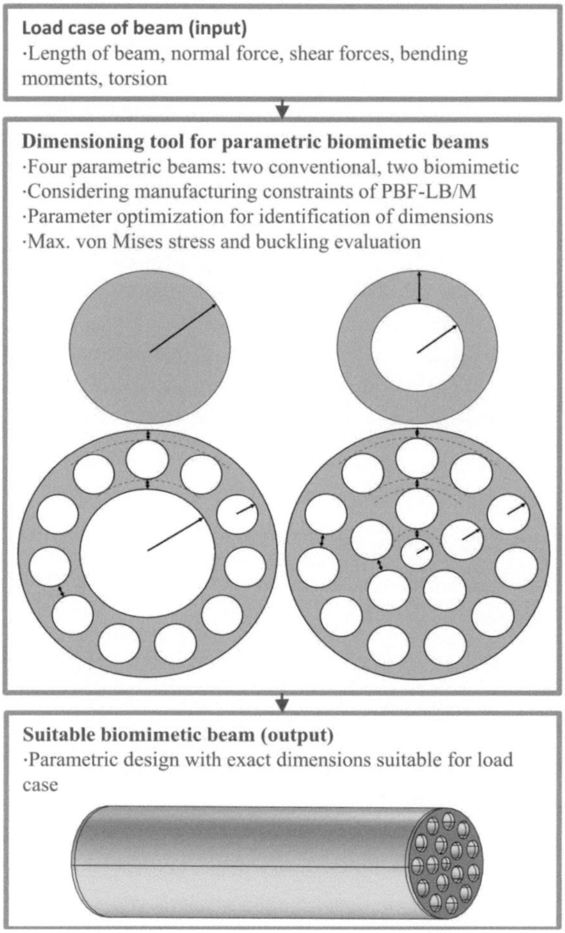

Figure 4.9. Schematic of the concept solution for dimensioning and selection of biomimetic beams. Modified and reproduced from [23] (licensed under CC BY 4.0) and [123] (licensed under CC BY 4.0).

4.7.2 Parametric beams

In this subsection, the development of parametric biomimetic beams is presented. In subsection 2.3.2, fundamentals of biomimetic beams were briefly presented. Based on the characteristics of the pseudo-stem of a banana plant, a great potential for the development of biomimetic beams with high bending stiffness and high buckling resistance was assumed. The full development of biomimetic abstractions from the biological role model and further development into parametric designs are presented in the following paragraphs.

In Figure 4.10 (a), a banana plant is shown and in Figure 4.10 (b), the cut surface of the pseudo-stem of a banana plant is depicted. Based on the biological role model, four initial beam cross section design were developed (see Figure 4.10 (c)-(f)). All abstractions consist of two or three cylindrical shells (corresponding to the epidermises of the pseudo-stem) and struts connecting the shells. Three designs with one cavity ring (Figure 4.10 (c)-(e)) and one design with two cavity rings were developed (Figure 4.10 (f)). It is noted that designs with three or more cavity rings were not developed as the manufacturing restriction of minimum distance between opposing walls would have led to rather large outer diameters. Such large outer diameters were expected to be of low relevance for the biomimetic design of PBF-LB/M components.

For the overall biomimetic design methodology, it was important that no support structures would be inside any of the beams, as removal of support structures from the inside of the beams would be hard or even impossible in the final component designs. Considering a threshold of 45 ° for the minimum downskin angle of AlSi10Mg components manufactured support-free in PBF-LB/M, critical areas are highlighted in Figure 4.10 (c)-(f). Considering a parametric design of the beam cross sections, the highlighted areas cannot be manufactured support-free for large parameter intervals. Therefore, the beam designs were considered not suitable for the biomimetic design methodology.

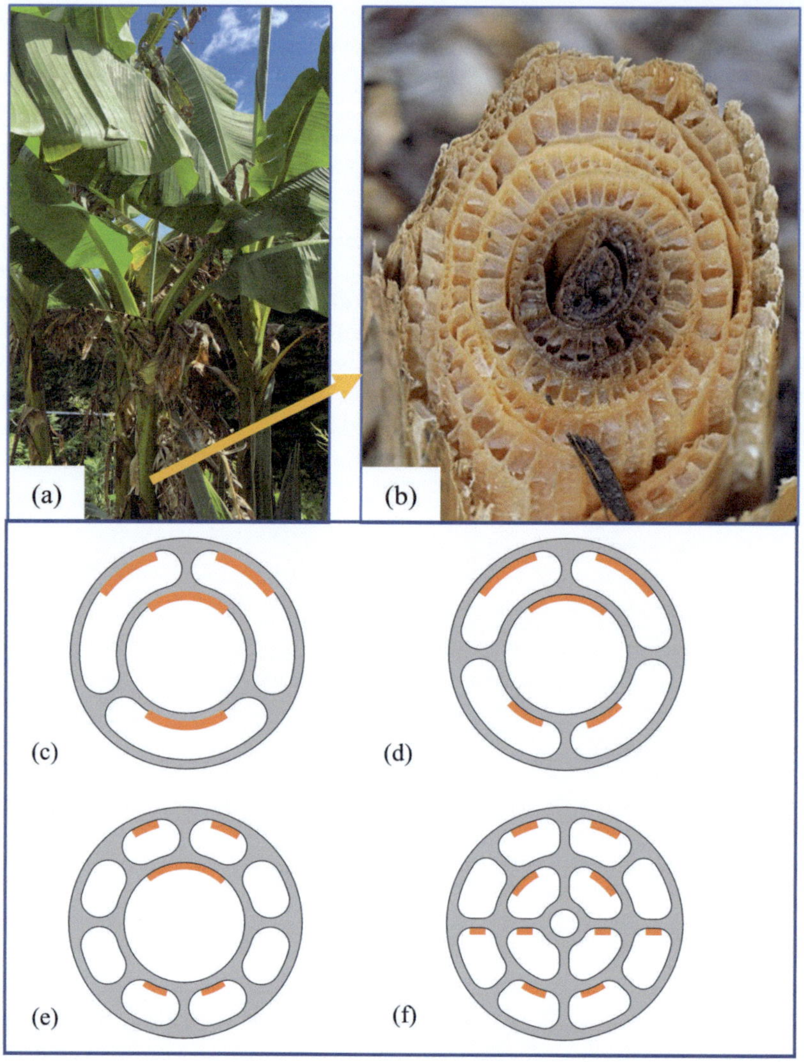

Figure 4.10. Generation of biomimetic beam designs based on pseudo-stem of a banana plant. (a): photograph of banana plants. © Röver, 2026; (b): photograph of the cut surface of the pseudo-stem of a banana plant. Reproduced with permission from [131]; (c)-(f): intermediate biomimetic abstractions of the banana plant pseudo-stem in the form of cross-sectional designs of beams. Problematic areas considering overhang manufacturing constraints by PBF-LB/M are highlighted in orange. Reproduced from [25] (licensed under CC BY 4.0).

Based on the previously presented designs, new biomimetic beam designs were developed (see Figure 4.11 (c) and (d)). Within certain intervals of the design parameters, these beams can be manufactured in AlSi10Mg without internal supports in the PBF-LB/M process. This is due to the fact that horizontal bore holes can be manufactured support free if their nominal diameter is equal to or below 15 mm (see subsection 2.1.2). Based on their appearance, the newly developed biomimetic beam designs were named revolver drum beam 1 (RD1) and revolver drum beam 2 (RD2). Manufacturability in AlSi10Mg in the PBF-LB/M process was successfully demonstrated for selected parameter sets of RD1 and

RD2 (see subsection 4.7.8). As additional options to the biomimetic beams, a parametric solid cylindrical beam (SC) design (see Figure 4.11 (a)) and a parametric hollow cylindrical beam (HC) design (see Figure 4.11 (b)) were utilized in this thesis. The radius of any cavity in the parametric designs is greater than or equal to 1.5 mm to ensure powder removability. Furthermore, the radius of any cavity must be smaller or equal to 7.5 mm to ensure support-free manufacturing. Additionally, the minimum wall thickness (which is also the minimum distance between any cavities) must be equal or greater than 1 mm to reduce the chances of local buckling in thin walls. Additional information on suitable lower and upper bounds of the various design parameters of the four beam designs can be found in [25,122,123].

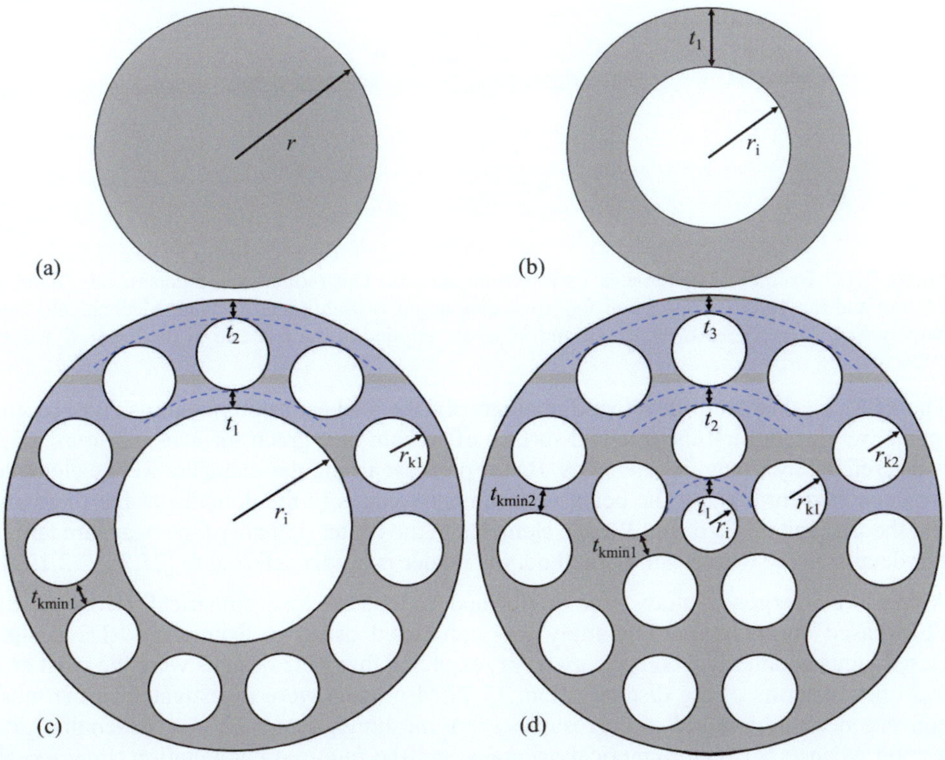

Figure 4.11. Conventional and biomimetic parametric beam designs. (a): solid cylindrical beam (SC); (b): hollow cylindrical beam (HC); (c): revolver drum beam 1 (RD1); (d): revolver drum beam 2 (RD2). Modified and reproduced from [123] (licensed under CC BY 4.0).

4.7.3 Parameter optimization

Parameter optimizations were performed in COMSOL 6.0 using its optimization module [55] and functions based on the popular Nelder-Mead method [132,133] and the theory of Lagrange multipliers. The objective was the minimization of the cross-sectional area of the beam. One constraint was given by the acceptable stress $\sigma_{acc,AlSi10Mg}$ which is 112.8 MPa (see section 3.3). Furthermore, the upper and lower bounds of the design parameters of the beam design were considered in the parameter optimizations.

An exemplary parameter optimization model is depicted in Figure 4.12. The model consists of three domains, one thin domain at the fixed end and one thin domain at the free end. Furthermore, there is the center domain covering the majority of the beam in which the stresses are evaluated during optimization. The two thin domains are excluded from the evaluation as discretization is expected to lead to physically meaningless stress peaks in these regions, that would compromise the results. Input loads for a parameter optimization are given by the axial force N_{PO}, the shear force along the first local axis $T_{1,PO}$, the shear force along the second local axis $T_{2,PO}$, the twisting moment $M_{t,PO}$, the bending moment around the first local axis $M_{1,PO}$, and the bending moment around the second local axis $M_{2,PO}$. These loads act at the free end of the beam.

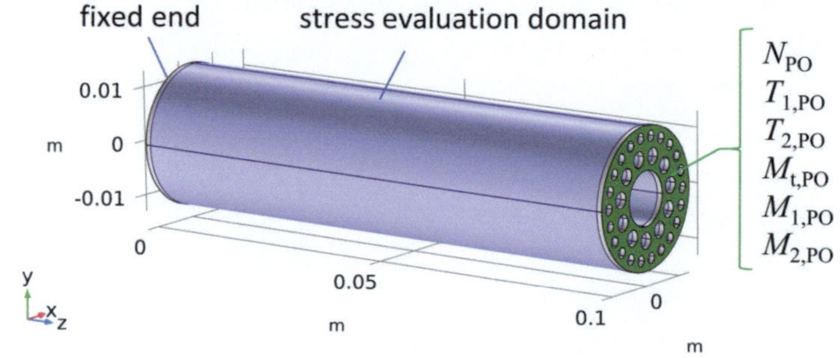

Figure 4.12. Exemplary parameter optimization model. The model was mechanically fixed at $z = 0$ m and the volume considered for stress evaluation is highlighted in blue. Mechanical loads, namely N_{PO}, $T_{1,PO}$, $T_{2,PO}$, $M_{t,PO}$, $M_{1,PO}$, and $M_{2,PO}$ are applied to the free end of the beam. © Röver, 2026.

The FEM meshes were developed analogously for all beam optimizations. Wedge elements were used. First the free end surface of a beam (see green surfaces in Figure 4.12) is discretized by triangular elements. Based on these triangular elements wedge elements are generated throughout the beam with thinner wedges in the domains at the fixed and end the free end of the beam. Wedge elements in the center domain of the beam are larger. For details on the discretization method, the reader is kindly referred to [25,26,122,123].

A mesh convergence study was performed to ensure low numerical errors in the FEM-based optimization. The study was conducted using an exemplary RD2 design. Based on its geometrical complexity it was expected that RD2 designs were the most critical ones in terms of the discretization. 17 FEM models were investigated in the mesh convergence study. Based on the results a set of meshing parameters was chosen that gave a good balance between numerical accuracy and the required computation time. Exemplary FEM meshes according to the final meshing sequence are depicted in Figure 4.13. For more detailed information on the mesh convergence study, the reader is kindly referred to [25] and [122].

4.7.4 Comparison of exemplary optimization results

Figure 4.13 shows optimization results for the four different beam design for an exemplary beam load case. All four beams were subjected to the same mechanical boundary conditions and have the same length. It can be seen that von Mises stresses in all four results are close to the maximum allowable stress $\sigma_{acc,AlSi10Mg}$ of 112.8 MPa. This characteristic suggests that close to no material can be removed from the beam design by variation of its design parameters without exceeding the allowable stress. It is a sign that the mass reduction objective seemed to have work well.

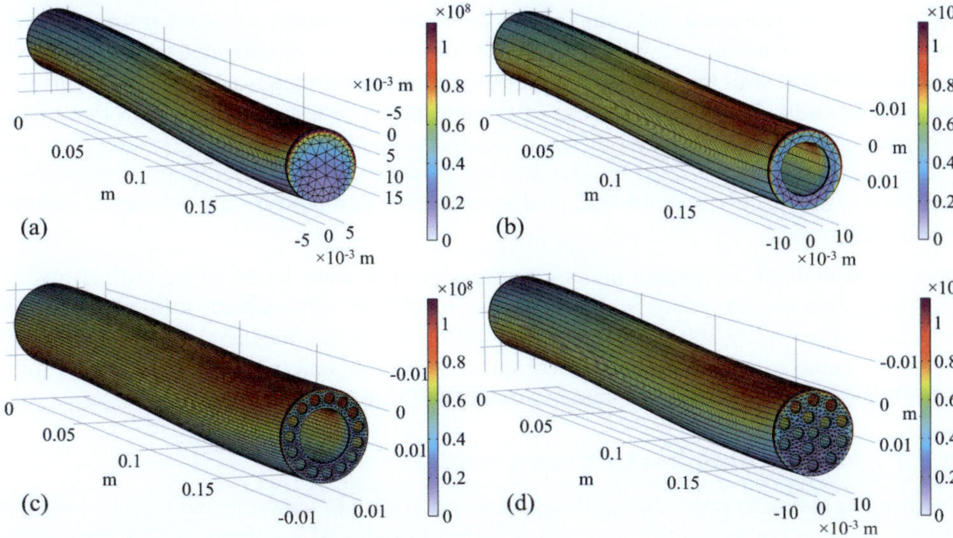

Figure 4.13. Exemplary FEM-based parameter optimization results of beams subject to the same mechanical boundary conditions. Deformation of the beams are exaggerated for improved visibility. (a): solid cylindrical beam (SC); (b): hollow cylindrical beam (HC); (c): revolver drum beam 1 (RD1); (d): revolver drum beam 2 (RD2). Modified and reproduced from [123] (licensed under CC BY 4.0).

4.7.5 Selection of best beam

In this step one optimized beam design has to be chosen for each beam of the auxiliary model for substitution. From the previous step, up to four optimized beam designs can be generated for each beam load case. The cross-sectional area of all beams that were optimized for a certain load case are compared and the beam with the smallest cross-sectional area is chosen. For details on this aspect, the reader is kindly referred to an exemplary spread sheet in [123].

4.7.6 Buckling analysis and re-dimensioning of beams

Exceeding the maximum allowable stress of the material is one possible failure criterion in mechanical component design. Another is given by buckling. All of the beams that were chosen in the previous step and that are under compressive loads have to be investigated with regard to buckling. Buckling analysis is performed in COMSOL 6.0 utilizing the same material parameters, maximum allowable stress, and boundary conditions. Furthermore, as in the case of the parameter optimizations a sufficiently fine finite element mesh has to be utilized. The output from the calculation is a critical load factor. A structure is critical with regard to buckling if this value is between 0 and 1 [130]. For values above 1, the structure is uncritical with regard to buckling.

Any beam that is uncritical with regard to buckling according to the analysis, is selected for the final design. If any beam is critical with regard to buckling, additional steps have to be performed. The outermost wall thickness parameter of the respective beam is increased by small increments until the design is no longer critical with regard to buckling. The outermost wall thickness parameters are given by parameter r of SC design, parameter t_1 for HC design, parameter t_2 for RD1 design, and parameter t_3 for RD2 design. Next, the cross section is evaluated and compared to the alternative beam solutions from the previous step. In case another beam is lighter, this beam must be investigated toward buckling and its outermost wall thickness parameter might need to be increased until it is not critical toward buckling. If the cross section of the alternative beam is smaller than that of the previous choice, the alternative beam is compared with the remaining beams from the previous solution of the step. In case another beam turns out to be lighter, this beam is analyzed and, if necessary modified, analogously. Following this routine, the most suitable beam is identified.

It is noted that for SC beams under compression that were found to be critical toward buckling, dimensioning may be done using an analytical approach according to the Euler buckling criterion [99,100]. A fixed-free condition might be considered. However, as most likely the beam is not purely under a compressive load, the newly developed design must be validated using the buckling analysis in COMSOL 6.0.

4.7.7 Interim biomimetic beams model

Performing the steps presented in this subsection allows to generate and select beams for the biomimetic component design. All optimized beams of a component are then introduced in a new 3D component design in COMSOL 6.0. It has to be ensured that the orientation of the beams is exact, especially for RD1 and RD2 beams, as they are not fully rotationally symmetric. Sheet-like elements are added to the component design in regions were supports are located and regions were loads act according to the original mechanical boundary conditions. Additionally, all beams are shortened such they do not intersect with each other and allow for a smooth node design. This interim biomimetic beams model is then exported in Standard for the Exchange of Product model data (STEP) file format and can be utilized in the next steps. Exemplary interim biomimetic beams models will be shown in section 4.9 (Figure 4.17 (c), Figure 4.18 (c), and Figure 4.19 (c)) together with exemplary models for other important steps of the complete biomimetic design methodology.

4.7.8 Additively manufactured beam demonstrators

To demonstrate the manufacturability of exemplary RD1 and RD2 beams, one PBF-LB/M build job was performed. AlSi10Mg powder with a particle size distribution of 20–63 μm was employed. The build job is depicted in Figure 4.14. RD1 and RD2 beams with exemplary design parameters were manufactured in build orientations of 0 °, 45 °, and 90 °. Data preparation w most important steps of the biomimetic design as performed in Netfabb. Manufacturing was performed via PBF-LB/M. A laser power of 170 W, a scan speed of 1200 mm/s and a hatch distance of 0.1 mm were used. The layer thickness was 20 μm, and the focus diameter of the laser was 70 μm. The build chamber was not preheated. It is noted that build job preparation and AM was performed by a technician of Hamburg University of Technology. It was demonstrated that the beams could be manufactured successfully without the need of any internal supports in the cavities of the beams. Furthermore, the beams did not show any visible defects or irregularities. Partly broken support structures indicate relevant residual stresses in both horizontally oriented beams. Preheating of the build chamber during AM could reduce residual stresses and prevent deformation and fracture of support structures in future build jobs of the beam structures.

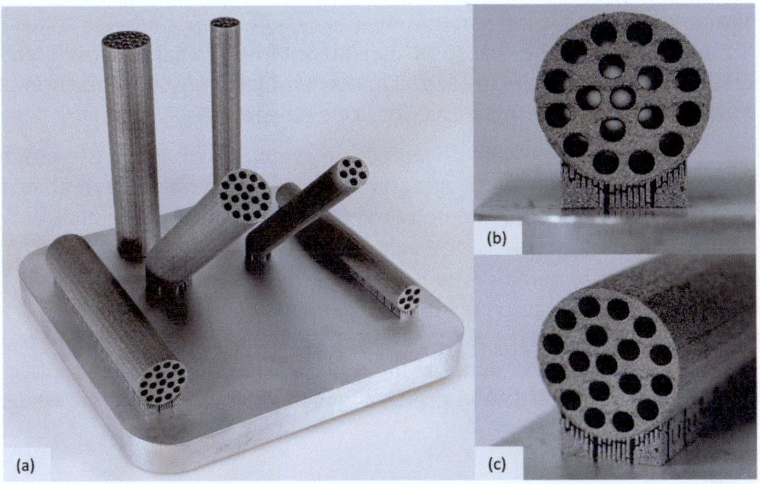

Figure 4.14. Exemplary revolver drum beams (RD1 and RD2) in three different build orientations (0 °, 45 °, and 90 °) manufactured by PBF-LB/M in AlSi10Mg. The build platform has a base area of 152 mm × 152 mm. (a): photograph of the entire build job; (b): close-up of the RD2 design manufactured at 0 °, showing that cavities are free of support structures; (c): additional close-up of the RD2 design manufactured at 0 °, showing no visible defects or geometric irregularities. Reproduced from [25] (licensed under CC BY 4.0).

4.8 Creation of nodes

In this section, the concept and methodology for the creation of nodes for the biomimetic component design are presented. It is noted that this section is largely based on two papers [23,26] that are part of this thesis.

The following goals were defined for the development of a methodology of node design in this thesis:

1. Ensure the successful transmission of forces and moments between the beams.
2. Ensure the structural integrity of the nodes and the overall component design.
3. Ensure the powder removability from the node as well as the cavities of the beams that are connected to the node.
4. Ensure that the node design and the resulting final component design consider the manufacturing restrictions of PBF-LB/M.
5. Ensure a small mass of the node.
6. Ensure that the design sequence of the nodes is robust and time-efficient.

In [23], the authors suggest the use of two different types of topology optimization approaches for the development of node designs in the biomimetic design process. However, topology optimization was not used for node design in the final biomimetic design methodology of this thesis due to the following considerations. Using topology optimization for node design has the advantage that light designs can be generated (goal five). Also goals one to four might be achieved using topology optimization approaches. However, topology optimization is rather expensive in terms of computational cost. This can lead to long calculation times for high resolution results. At the same time, the robustness of topology optimization approaches has to be considered low. To achieve high quality results often experienced engineers are needed. This is why topology optimization approaches were not further considered in this thesis for node design.

Instead a design methodology was developed in which outer shells are combined with lattice infill structures. Certain lattice structures were found to have favorable characteristics for efficient powder removal and demonstrated considerable potential for lightweight design [134,135]. Therefore, the use of lattice structures was chosen for a considerable share of the volume of the nodes. In addition, an outer shell was selected as part of the node design. The outer shell prevents high stresses in the outer areas of the nodes as the continuous geometry prevents the notch effect, that was expected to lead to problems if a lattice structure was in the outer areas of the nodes. Additionally, many lattice structures are self-supporting in the PBF-LB/M process, which is beneficial for the design methodology.

4.8.1 Solid volume nodes

In this subsection, the design methodology for the solid volume nodes is presented. Inventor 2024 was used for this part of the methodology. The STEP file of the interim biomimetic beams model is used as an input for this solid volume node generation. Relevant steps of the methodology for an exemplary node are depicted in Figure 4.15. The design methodology can be performed for one node at a time.

In a first step, 2 mm offsets are added to the outer radii of all beam ends that are to be connected by the node (Figure 4.15 (a)). In the next step, the loft function is used to generate solid connections between the offset circles (Figure 4.15 (b)-(d)). In case any hole is present in the center of the node after all connections were made, an extrusion is added to the center part of the node manually, to close it. Next, the center areas of the nodes are smoothed utilizing the fillet function (Figure 4.15 (e)). Then a smooth transition toward the center of the beams is added (Figure 4.15 (f)), to prevent the stress peaks at the interfaces between node and beams. This step adds additional material to the beams. Next, the node is separated from the beams by plane cuts ((Figure 4.15 (h)) and then extrusions are added to the nodes according to the cavities of the respective beams (Figure 4.15 (i)). The depths of the extrusions are at least equal to the unit cell size of the lattice structure that is to be used for node design. After all nodes are designed a single component consisting of the beams and nodes is exported as a STEP file.

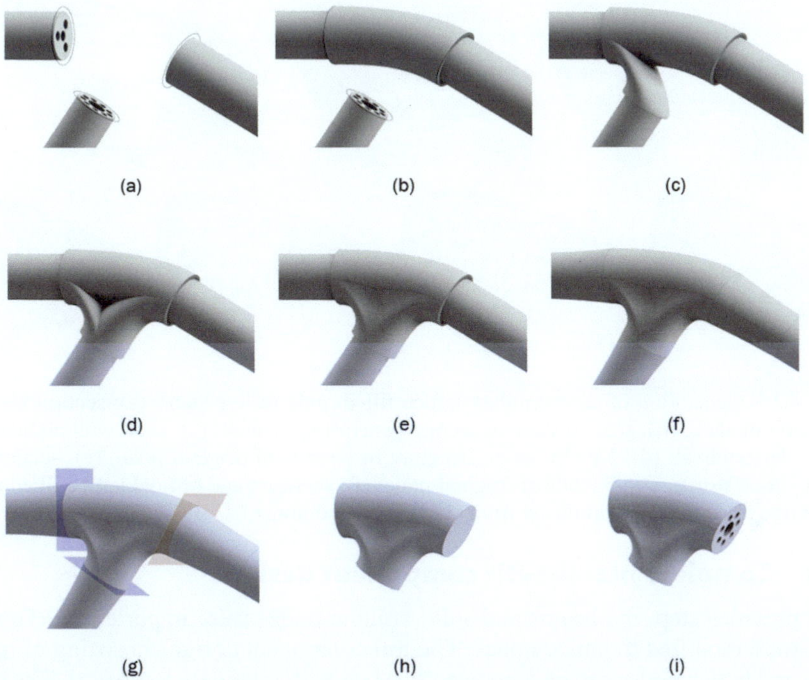

Figure 4.15. Generation of an exemplary solid volume node in five steps. (a): added offsets; (b)-(d): connection of all beams; (e) and (f): smoothing; (g) and (h): separation of node; (i): extrusion of cavities into nodes. Reproduced from [123] (licensed under CC BY 4.0).

4.8.2 Integration of a lattice infill structure

For the generation of the final node designs by integration of the lattice infill structure, nTop was used. The STEP file as exported from Inventor 2024 is imported to nTop and the beams and nodes are converted to implicit bodies. Then all nodes are modified according to the following steps: First a shell is generated from the previous node design (Figure 4.16 (a)). Wall thickness values of 3 mm were found to be suitable for the components investigated within this thesis. However, adaptation of this parameter may be necessary for other components based on the overall dimensions of the component and the specific node. Next, perforations are added (Figure 4.16 (b)) and at least one cut-out is performed to improve powder-removability from the component (Figure 4.16 (c)). With this step the shell part of the node is complete.

Next, the original solid volume node is taken once more as an input and is fully filled with a lattice structure (Figure 4.16 (d) and (e)). Within this thesis, the Kelvin cell was found to be beneficial with regard to mechanical integrity as well as powder removability. Unit cells of the lattice structure had edge lengths in the range of 4 mm to 5 mm and wall thickness values were between 2 mm and 2.5 mm. In the final step, the lattice structure is combined with a Boolean union operation with the shell part to obtain the full node design (Figure 4.16 (f) and (g)).

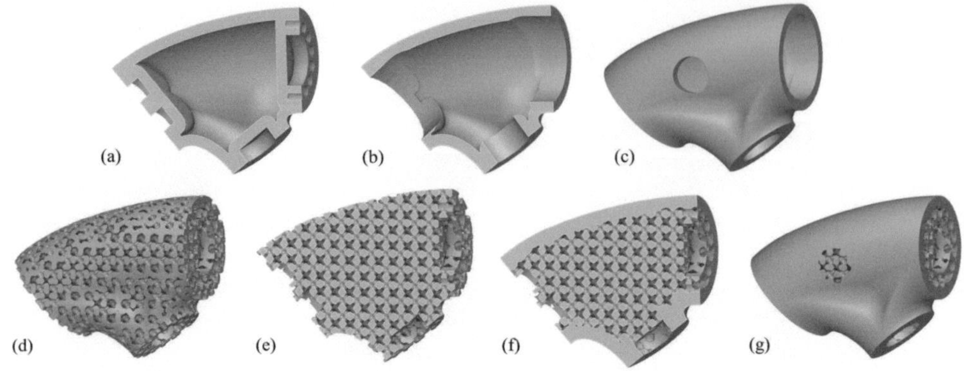

Figure 4.16. Generation of an exemplary lattice filled node in five steps. (a): section view of first generation of shell; (b): section view of second generation of shell; (c): shell with circular cut outs for powder removal; (d): Kelvin lattice structure in bounds of original node; (e): section view of Kelvin lattice structure in bounds of original node; (f): section view of final lattice filled node; (g): final lattice filled node. Reproduced from [123] (licensed under CC BY 4.0).

4.8.3 Complete biomimetic component design

In the previous step, the beams and solid volume nodes were imported to nTop and the nodes were modified to lattice nodes. The full component design consisting of optimized beams and lattice nodes can be generated in nTop with relatively low effort. The complete component designs can be exported from the software in STL file format as well as STEP file format. For more details on the matter of node design, the reader is kindly referred to [26] and [123].

4.9 Application of biomimetic design methodology to three topology optimization results

In this section, the results from application of the previously presented design methodologies to the three exemplary topology optimization results, namely Michell cantilever, L-shape, and MBB beam are presented. It is noted that this section is largely based on [26] which is part of this thesis.

In Figure 4.17, Figure 4.18, and Figure 4.19, the most important steps of the biomimetic design methodology for the exemplary components can be found. Panels (a) of these figures show the geometrically upscaled topology optimization results, where the acting forces were also upscaled as presented in section 4.4. The panels labeled (b) depict the geometrically upscaled auxiliary 3D models that could be derived based on the methods presented in section 4.5. Results from the application of the methods presented in sections 4.6 and 4.7 can be found in panels (c). The final biomimetic component designs according to section 4.8 are depicted in panels (d) and (e) of the three figures. In the following subsections, additional information on the application of the design methods to the three exemplary components is given.

4.9.1 L-shape

The biomimetic L-shape consists of seven SC beams. For nodes connecting SC beams, a modified design approach for the nodes was applied, as no powder removal from the SC beams was necessary. In the modified node design, no 2 mm offset in the solid volume node generation was done. Further, no lattice infill structure was utilized in regions of the nodes that connect SC beams. For nodes that connect exclusively SC beams, no lattice infill was used. Nodes that connected SC beams with beams that had cavities were only partly filled with the lattice structure which ensured powder removability from any cavity of the beams. A benefit of the modified node design is the improved stress distribution at the interface of the nodes and the SC beams.

Three beams of the L-shape were found to be critical with regard to buckling. All three beams were SC beams. The Euler buckling criterion was used for all three beams for re-dimensioning. Buckling analysis of the re-dimensioned beams yielded that they were not critical toward buckling. As they were still lighter than their HC, RD1, and RD2 counterparts, the updated SC beams were used for the final design of the biomimetic L-shape.

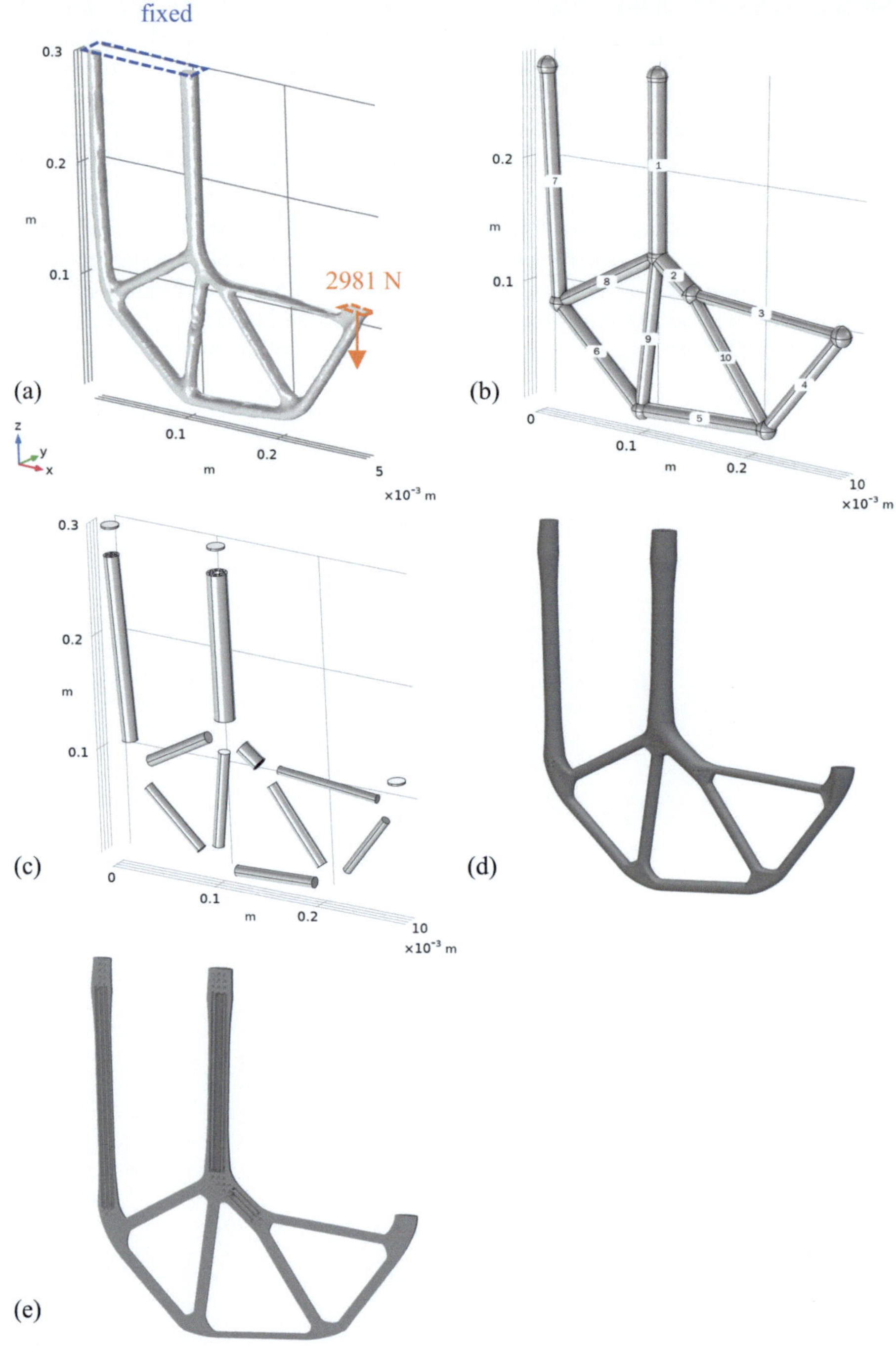

Figure 4.17. Views of reference topology optimization result, interim results, and final component design of the L-shape. (a): topology optimization result with boundary conditions; (b): abstraction

of topology optimization result; (c): arrangement of biomimetic beams; (d): final biomimetic component design; (e): section view of final biomimetic component design. Modified and reproduced from [123] (licensed under CC BY 4.0).

4.9.2 Michell cantilever

None of the optimized beams of the Michell cantilever were critical with regard to buckling. Therefore, no re-dimensioning of any optimized beam of this component was necessary. Furthermore, one node was designed using another explorative design approach. The node connects one SC and three HC beams. A solid volume node was designed using Inventor 2024 and then the loft function was used to create a powder outlet for the cavities of the HC beams. No lattice structure was integrated into this node.

To reduce computational efforts in the study, optimizations of SC, HC and RD2 beams were conducted for selected beam load cases only. Furthermore, for beams #15 and #16 the RD2 beam was preselected to reduce computational efforts.

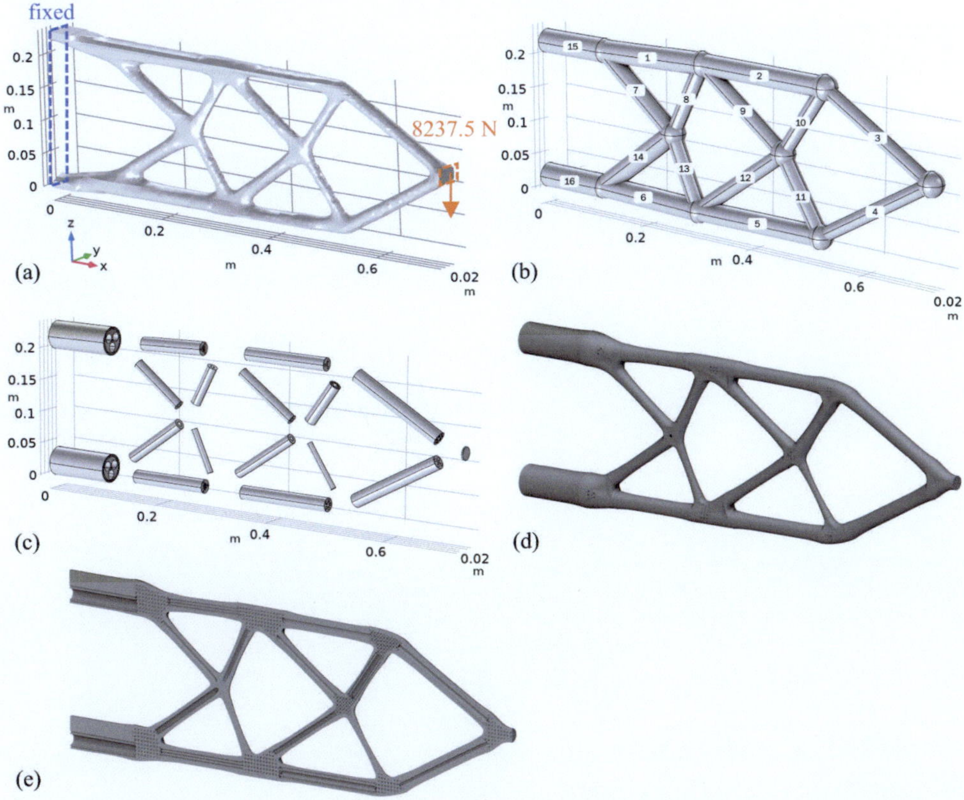

Figure 4.18. Views of reference topology optimization result, interim results, and final component design of the Michell cantilever. (a): topology optimization result with boundary conditions; (b): abstraction of topology optimization result; (c): arrangement of biomimetic beams; (d): final biomimetic component design; (e): section view of final biomimetic component design. Modified and reproduced from [123] (licensed under CC BY 4.0).

4.9.3 MBB beam

In the generation of the biomimetic design of the MBB beam, beam #2 was omitted in the final design, as it was relatively short. The two nodes that connected beam #2 were integrated into a single node. Also, a sheet element was added to this node to which the force could be applied. In the case of the MBB beam, none of the optimized beams were found to be critical with regard to buckling. Therefore, no re-dimensioning was necessary.

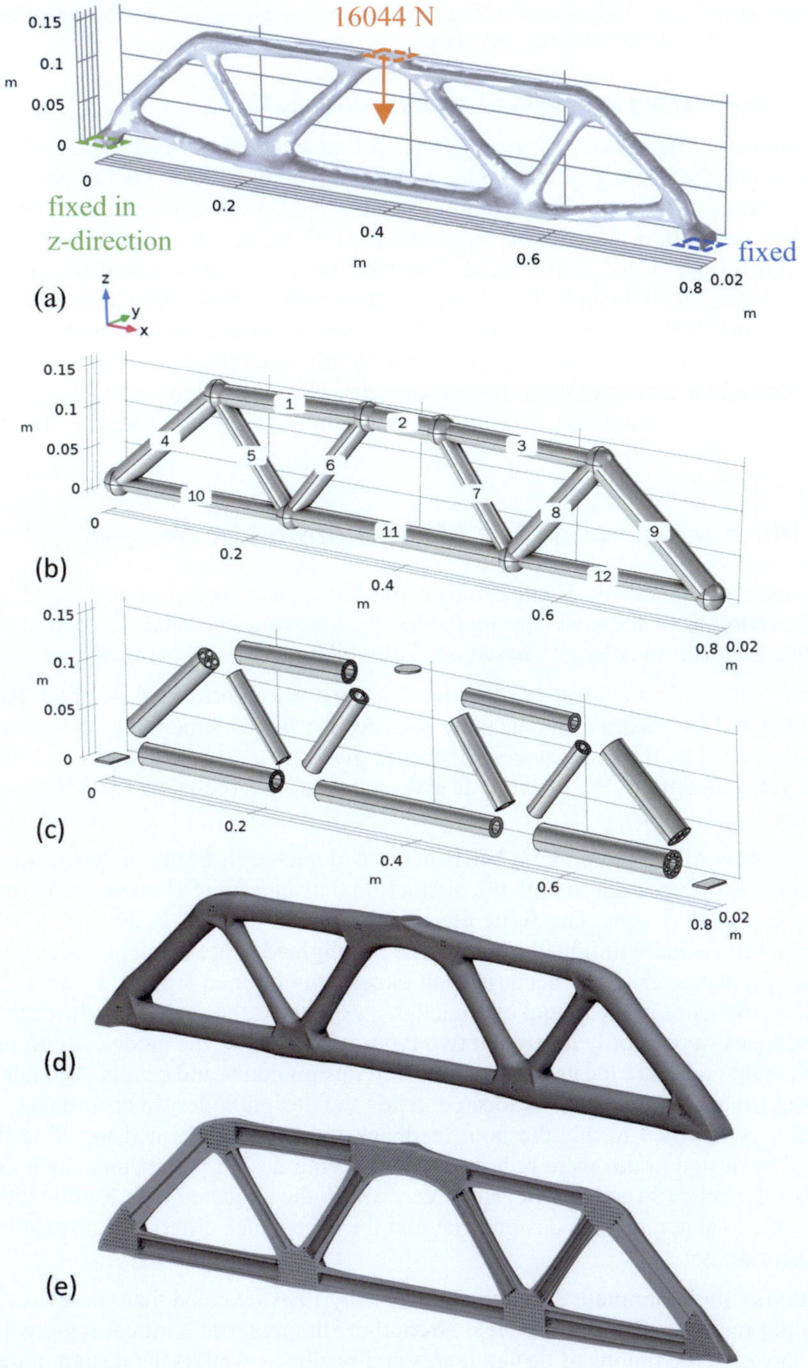

Figure 4.19. Views of reference topology optimization result, interim results, and final component design of the MBB beam. (a): topology optimization result with boundary conditions; (b): abstraction of topology optimization result; (c): arrangement of biomimetic beams; (d): final biomimetic

component design; (e): section view of final biomimetic component design. Modified and reproduced from [123] (licensed under CC BY 4.0).

4.9.4 Qualitative analysis of component designs

The biomimetic Michell cantilever contains all four kinds of beams considered in the design methodology, namely SC, HC, RD1 and RD2. It contains multiple lattice-filled nodes and one node that was designed using an explorative design approach with one powder outlet (see subsection 4.9.2). The biomimetic MBB beam only features HC and RD1 beams. All its nodes are lattice filled. The biomimetic L-shape contains numerous SC beams and three RD1 beams. The design contains three solid nodes without any lattice infill, three nodes that are partly filled with a lattice structure, and two nodes that are entirely filled with a lattice structure. All three biomimetic component designs closely resemble the initial topology optimization designs. However, they show significantly increased geometric complexity due to the integration of optimized beams and modified nodes.

4.10 Mechanical validation of the biomimetic designs

In the next step, mechanical validation of the biomimetic component designs was performed according to the five-step methodology that was presented in section 4.3. It is noted that this section is largely based on [26] which is part of this thesis.

Stationary mechanical analyses of the three designs were performed in nTop. Automatically generated tetrahedra elements were used for the finite element mesh. Analyses were performed based on the acceptance criterion as given by $\sigma_{acc,AlSi10Mg}$ (equal to 112.8 MPa) and the yield strength of the material as given by $\sigma_{y,AlSi10Mg}$ (equal to 141 MPa) according to section 3.3.

A section view of the biomimetic MBB beam is depicted in Figure 4.20 together with a detail view. It is important to note the distinction between finite element nodes and nodes of the component design. The finite element mesh can be seen in the figure as well as finite element nodes highlighted in blue. The highlighted finite element nodes do not satisfy the acceptance criterion because they exceed the defined stress threshold. Multiple finite element nodes in the model exceeded $\sigma_{acc,AlSi10Mg}$. At the same time the critical finite element nodes were mainly located in two types of regions of the model. Multiple critical finite element nodes are located at the unions of design nodes and beams. A small number of critical finite element nodes is located inside the design nodes. In accordance with the methodology outlined in this thesis, a feedback loop was performed to refine the node design. The design nodes were redesigned by varying design parameters, such as the selection of the lattice structure type. However, within the allocated time for this task, it was not possible to generate node designs that met the acceptance criteria for all regions of the component design.

Validation of the biomimetic Michell cantilever by FEA revealed that von Mises stresses in multiple regions exceeded the yield strength of the material. Critical regions included also in this case the unions of design nodes and beams as well as the design nodes themselves including the lattice structure. Additionally, in this model some critical finite element nodes were present in regions close to where forces were applied and in regions close to mechanical supports. The explorative node design approach connecting one SC and three HC beams was successful, as no critical finite element nodes were present at this design node. FEA of the biomimetic L-shape yielded that von Mises stresses exceeded the yield strength of the material at unions of design nodes and beams. Furthermore, critical finite element nodes were present at certain design nodes including those design nodes that were developed with the alternative node design method for SC beams. Also, for these two biomimetic component designs, feedback loops to redesign of the nodes could not achieve satisfying node designs within the time allocated to this task.

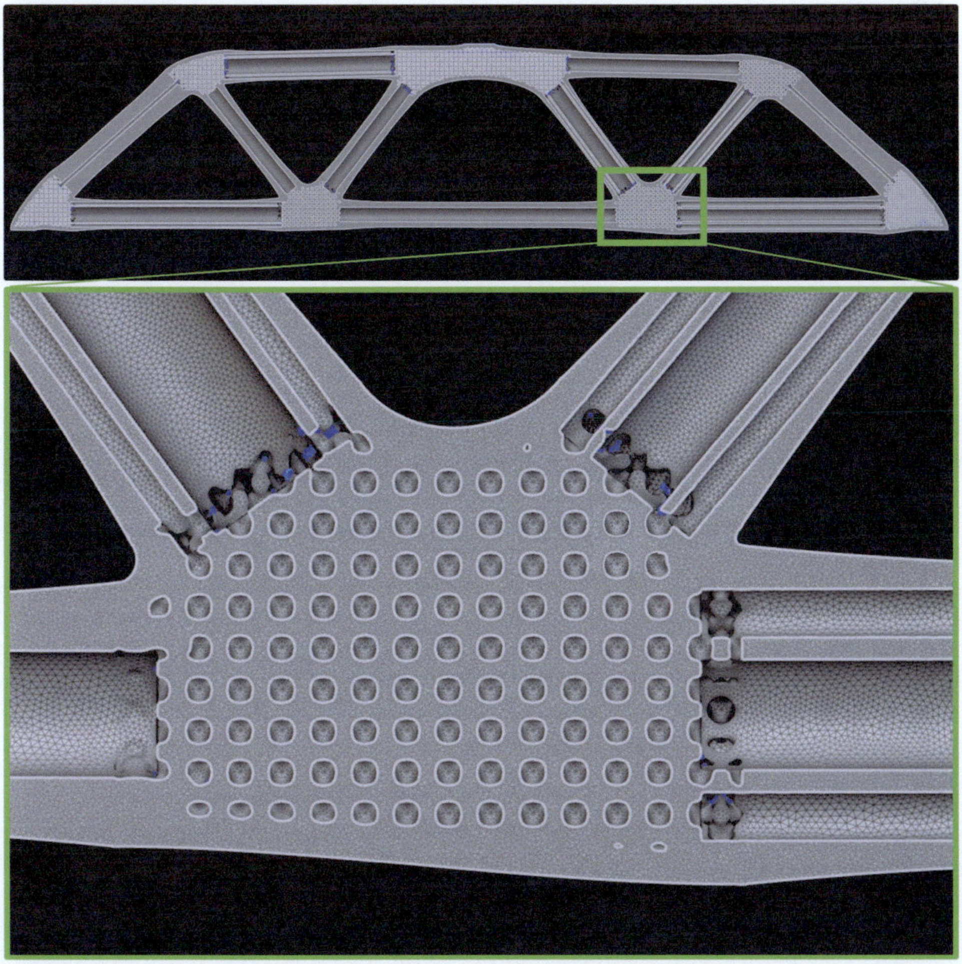

Figure 4.20. Section view and detail view of the final MBB beam biomimetic component design. Finite element nodes in which the von Mises stresses exceed 80 % of the yield strength of the material (112.8 MPa) are highlighted in blue. Modified and reproduced from [123] (licensed under CC BY 4.0).

4.11 Assessment of the lightweight performance

After FEA, an evaluation of the masses of the components was performed according to the five-step methodology that was presented in section 4.3. It is noted that this section is largely based on [26] which is part of this thesis. The evaluation of mass was performed based on the volumetric component designs. The topology optimization designs (Figure

4.19 (a)-(c)) were compared with their biomimetic counterparts (Figure 4.19 (j)-(l)). In Table 4.2 the masses of the three upscaled topology optimization component designs can be found as well as the masses of the three biomimetic component designs. Comparing topology optimization designs with their biomimetic counterparts, biomimetic designs achieved a significant reduction in mass. Weight reduction in case of the Michell cantilever was 30.3 % and weight reduction in case of the MBB beam was 14.9 %. Application of the biomimetic design methodology to the L-shape reduced the mass by 12.5 % compared to the topology optimization design.

Table 4.2. Comparison of the masses of topology optimization designs and biomimetic component designs. Reproduced from [123] (licensed under CC BY 4.0).

Component design	Mass [g]	Weight reduction compared to topology optimization result
Michell cantilever – topology optimization	3069.3	-
Michell cantilever - biomimetic	2139.7	30.3 %
MBB beam - topology optimization	2220.4	-
MBB beam - biomimetic	1889	14.9 %
L-shape - topology optimization	384.6	-
L-shape - biomimetic	336.5	12.5 %

4.12 Additively manufactured biomimetic design demonstrator

A demonstrator of the biomimetic component design was manufactured. It is depicted in Figure 4.21. The Michell cantilever biomimetic design was considered. A down-scaled section of the component design was manufactured. The section was scaled by a factor of 0.5. Furthermore, the nodes were redesigned using the original unit cell size of 5 mm instead of being down-scaled to ensure powder removability. Furthermore, cut-outs were introduced into the component design to make visible the internal structures in the various beams and nodes. The demonstrator contains all four types of beams (SC, HC, RD1, and RD2), modified Kelvin cell nodes, and an alternative node design. As in the case of the beam demonstrators (see subsection 4.7.8), data preparation was performed in Netfabb and manufacturing was done via PBF-LB/M using a laser power of 170 W, a scan speed of 1200 mm/s, a hatch distance of 0.1 mm, a layer thickness of 20 μm, and a 70 μm focus diameter of the laser. The build chamber was not preheated. For this build job AlSi10Mg powder with a particle size distribution of 20–63 μm was employed. It is noted that AM was performed by a technician of Hamburg University of Technology.

Manufacturing was successful and the final component did not show any visible defects, except for one. A small irregularity was found at the cut-out of the center node. The cut-out resulted in a relatively thin-walled unsupported region. The small wall thickness in the region and subsequent heat accumulation or deformation by the recoater in the PBF-LB/M process are possible reasons for the irregularity. The use of support structures in this region can be expected to prevent this problem in future build jobs.

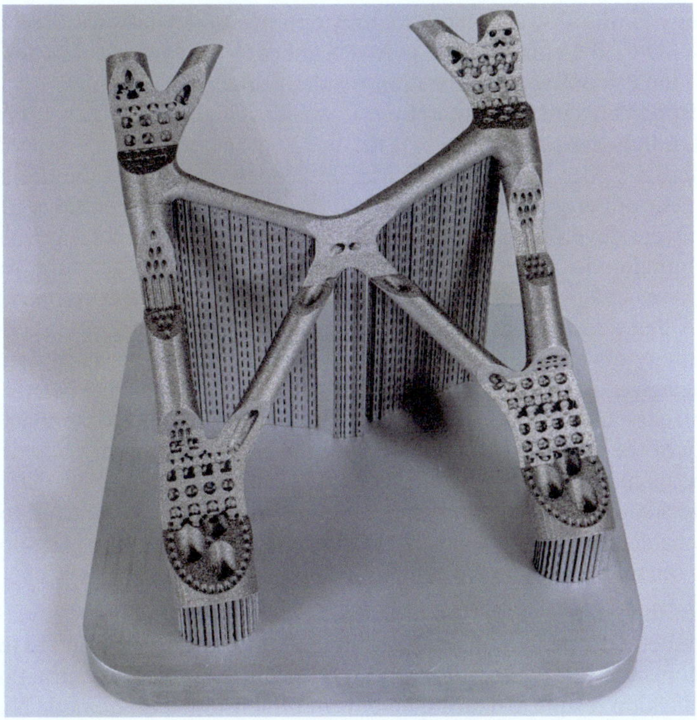

Figure 4.21. Biomimetic design demonstrator that was manufactured by PBF-LB/M in AlSi10Mg. The build platform has a base area of 152 mm × 152 mm. © Röver, 2026.

4.13 Discussion on the biomimetic design methodology

Within this thesis, a methodology for biomimetic design of structural components manufactured via PBF-LB/M was developed and assessed. The primary research question was how to develop a biomimetic design methodology to redesign mechanical components generated from density-based topology optimization by substituting cylindrical beams with biomimetic beams, while considering manufacturability by PBF-LB/M. The secondary research question was how biomimetic designs generated using such a methodology compare to the original topology optimization designs in terms of their lightweight characteristics. It is noted that this section is largely based on [23–26] that are part of this thesis.

4.13.1 Synopsis of the principal discoveries

The key outputs of this thesis are summarized as follows. The development of the biomimetic design methodology was performed based on the V-model. Based on an initial concept solution a more detailed five-step design methodology for biomimetic components was developed. The methodology has a density-based topology optimization result as an input and has a biomimetic component design as an output. Within the design methodology, the manufacturing restrictions of the PBF-LB/M process are considered. The biomimetic component consists of biomimetic beams and adapted nodes. For each of the five steps, detailed submethodologies were developed and assessed.

A methodology for the abstraction of topology optimization results was developed. Within the method a 3D model consisting of cylindrical beams and spherical nodes is developed based on the input topology optimization result. The used skeletonization approach was based on existing concepts from literature. It was developed and afterwards implemented in COMSOL 6.0. For the second step of the five-step methodology, a procedure for evaluation of internal forces and moments was developed that allows the definition of load cases for dimensioning of biomimetic beams. For the third step, solutions for biomimetic beam design were developed. Two parametric conventional beam designs as well as two parametric biomimetic beam designs considering the manufacturing restrictions of PBF-LB/M were developed. Parameter optimization models were developed for these four beam designs that allow the dimensioning of beams with a focus on lightweight design. The methodology developed for this step also considers buckling analysis of the beams, potential redesign of the beams, and the selection of the most suitable beams for a component. Exemplary beam designs were manufactured by PBF-LB/M to demonstrate the manufacturability of the designs. For the fourth step, a methodology for node design was developed. The submethod includes solid volume node design that gives the general shape of a node and a subsequent step in which a lattice infill structure is added to increase the lightweight character of the node and to ensure powder removability. The nodes combine the beams to obtain a complete biomimetic component design.

In step five of the methodology, the biomimetic design methodology is validated. Combining all submethods, the biomimetic design methodology was applied to three common topology optimization designs. Interim design results namely, 3D abstractions of the topology optimization results and arrangements of dimensioned biomimetic beams could be generated successfully. Furthermore, full biomimetic component designs were generated. The designs consist of conventional and biomimetic beams and lattice-filled nodes. The biomimetic component designs were analyzed by FEA and their masses were evaluated. Numerical results yielded that there are critical regions of high stress in all three of the biomimetic component designs. All critical finite element nodes have been associated with specific regions within the biomimetic designs. Mass evaluation showed that the biomimetic component designs are 12.5 % - 30.3 % lighter compared to their topology optimization counterparts.

The biomimetic design methodology allows similar component designs to be produced by engineers with different backgrounds. The detailed methods ensure that qualitatively similar and high-quality designs are generated. Furthermore, the methodology enables a traceable component development process, as interim results can be saved, and the methodology itself serves as the documentation needed to understand the steps taken between interim results.

4.13.2 Interpretation of the findings in context of existing research

The developed submethodology for abstraction of topology optimization results generates auxiliary 3D models. The method is based on the potential field approach for skeletonization (see subsection 2.2.2). The abstraction of 3D mechanical topology optimization results by spherical nodes and cylindrical beams using skeletonization was proposed in previous studies [76,85]. The submethodology that was developed in this thesis was implemented in COMSOL 6.0. It combines the benefits of the approaches from literature with that of the used software: Potential field approaches for skeletonization are known for low sensitivity to boundary noise and tend to generate smooth curves [79] which allowed for the generation of high-quality abstractions of the topology optimization designs by lines and points in this thesis. Building on these results, the generation of auxiliary 3D abstractions that consist of cylindrical beams and spherical nodes paves the way for the next step of the biomimetic design method, namely evaluation of internal forces and moments in the 3D abstractions. The evaluation is done by FEA in COMSOL 6.0 as this software offers a function that is well-suited for this purpose. Therefore, the developed methods provide new solutions contributing to the fields of skeletonization and 3D abstraction while being tailored to integrate well into the more comprehensive biomimetic design methodology presented in this thesis.

Comparing the biomimetic design methodology of this thesis with other methods on redesign of topology optimization results [76,85] it stands out as it includes evaluation of internal forces and moments and subsequent lightweight optimization for each beam of the design. In other studies such as [76], dimensioning of beams is done based on geometric evaluation of the topology optimization and redesign of a beam is only done if found necessary during FEA.

The generation of two novel and true biomimetic parametric beam designs that consider the manufacturing restrictions of PBF-LB/M and that are based on the pseudo-stem of the banana plant extends the research field. The development of parameter optimization models of the two biomimetic and two conventional beam designs that are tailored to be used in the overall biomimetic design methodology are additional contributions of this thesis.

Reflecting on biomimicry, the thesis combines topology optimization and lattice structures that are considered biomimetic according to [95] with beam designs that are considered true biomimetic according to [95] as they were developed using a direct input from nature. Therefore, the methodology facilitates the wider adoption of true biomimetics in DfAM by reducing barriers, addressing the scarcity of such designs in the field [95].

Considering the complete biomimetic design methodology that was presented in this thesis, the five-step design methodology was inspired by the integrated product development process for biomimetic optimization [93,94] that is rather coarse-grained. The new five-step methodology provides a detailed procedure for the redesign of topology optimization designs by integration of biomimetic beams and adapted nodes. Thus, it focuses on mechanical components and one specific approach to generate detailed designs of components. At the same time, it closes the gap between topology optimization results and the final biomimetic component design that can be easily edited if necessary. The parametric beam designs and the submethodology for node design allow for an easy adaptability of the resulting biomimetic component design. This aspect is of great importance as the conversion of topology optimization results to easily-editable design files remains a challenging step [76] that is overcome within the biomimetic design methodology of this thesis.

Application of the detailed five-step biomimetic design methodology to three input topology optimization designs demonstrated the successful integration of biomimetic beams into optimized structures. This research illustrates that the methodology is applicable to multiple inputs, offering a more general yet detailed approach. In contrast, previously existing studies either enhanced unique components using specialized design procedures or provided only coarse-grained design methods [10,93,94]. Given the fact that topology optimization designs are mathematical optima, improvement of such designs is a challenging task. The weight reductions of 12.5 % - 30.3 % of the biomimetic designs compared to the original topology optimization designs that could be shown in this thesis, prove the effectiveness of the presented design methodology. Consequently, the biomimetic design method proposed in this thesis substantially advances the field and partly closes the research gap for biomimetic design tools identified in the scientific review article from 2020 [11].

4.13.3 Implications

In this subsection, the most important theoretical and practical implications of the work are discussed. Beginning with the accessibility of the results, it is noteworthy that all four research articles included in this cumulative thesis have been published as open access. Furthermore, pertinent research data providing exemplary optimization models, code, and component design files for all important steps of the methodology have also been made available as open access. This significantly lowers the barrier for developing biomimetic mechanical components using the presented methodology, allowing anyone to build directly upon this work. Additionally, some of the developed submethodologies have been automated through code, further enhancing the accessibility and ease of use of the methodology.

Regarding its relevance for DfAM, the results in this thesis suggest that the design methodology can be improved such that it improves a majority of truss-like mechanical topology optimization results by generating biomimetic component designs with reduced mass while ensuring structural integrity. This advancement could establish the proposed biomimetic design methodology as a standard step following topology optimization for any mechanical topology design intended for production via PBF-LB/M. This is pertinent for engineers and decision-makers involved in the design of mechanical lightweight components manufactured using PBF-LB/M.

The methodology could be applied across various lightweight industries and their applications to explore the potential for further mass reduction in components. Exemplary applications that were introduced in the introduction of this thesis which could be improved by the methodology include satellite brackets, space rocket launcher brackets, and civil aviation brackets [12–18] as well as automotive vehicle structures and suspensions [19–21], and bicycle components such as frames and forks [22].

Furthermore, the methodology could be applied to components that are designed to be manufactured by powder bed fusion of polymer with a laser beam (PBF-LB/P). As PBF-LB/M and PBF-LB/P are very similar processes, it is expected that the biomimetic design methodology can be used to develop mechanical PBF-LB/P components with very few adaptations. The most important adaptation is that the polymer material parameters have to be used throughout all design steps. One benefit of PBF-LB/P is that the process typically does not require support structures. This reduces the restrictions in design steps such as beam optimization and node design compared to the case of PBF-LB/M components. Potential applications are Unmanned Aerial Vehicle (UAV) airframe structures that are truss-like topology optimization designs and that are manufactured using polymers [136–138]. Apart from PBF-LB processes, the methodology could be applied in modified versions to components that are designed for other AM processes such as powder bed fusion of metal with electron beam (PBF-EB/M), directed energy deposition (DED), binder jetting (BJT), material extrusion (MEX), sheet lamination (SHL), and vat photopolymerization (VPP).

Another possible application of the biomimetic design approach presented in this thesis lies in architecture and civil engineering. Mechanical topology optimization and algorithm-aided design tools are used to generate truss-like designs in the field [139–142] and integration of biomimetic beams could improve them. Also, tree-like structures are used in architecture and civil engineering [143] which could benefit from the integration of biomimetic beams.

4.13.4 Time, cost, quality, and sustainability

In this subsection, the implications of the methodology with regard to time, cost, quality, and sustainability are discussed in more detail. They are considered for the case that the methodology is used for lightweight design of PBF-LB/M components in the aerospace industry. Starting with the weight reduction potential, additional weight savings in numerous mechanical components in the range between 10 % and 30 % compared to topology optimization component designs would have a direct effect on the aircraft's weight. Additionally, reduced expected weight of structural components has further implications in the design of an aircraft as mass is one of the main factors considered in the preliminary design phase of an aircraft [144] and thus has implications for numerous of its systems. Reduced weight influences the design or choice of systems such as an aircraft's engines [144] or its landing gear [145] as well as others [144], potentially allowing for lighter options. The weight reduction of the aircraft through direct and indirect effects has an influence on multiple aspects: it leads to reduced fuel consumption during operation [145], which reduces operating costs as well as decreases CO_2 emissions [144] and, therefore, improves sustainability. Additionally, lighter aircraft tend to be quieter [146], further improving environmental sustainability and quality aspects such as passenger comfort and contributing to the aircraft meeting noise certification limits.

Another benefit in terms of cost and sustainability is that less material is necessary in manufacturing, due to the reduced volume of the component design that is directly linked to its reduced weight [147,148]. This reduction in volume also leads to decreased time needed during additive manufacturing, which is beneficial in terms of cost and lead time [147].

Other important implications are given by the fact that the digital value chain was optimized by the biomimetic design methodology presented in this thesis. The methodology integrates multiple aspects. As the methodology is very detailed and partly automated, less experienced engineers are sufficient to generate biomimetic lightweight designs using it. Furthermore, automated submethodologies save time in the design process. Both of these aspects are connected to reduced personnel costs as well as a reduced lead time. Furthermore, the methodology includes the step of conversion of a topology optimization result into an easily-editable file. This is important in subsequent steps such as design iterations and planning of computer-aided machining of the component [149], which therefore reduces personnel costs and lead time. Furthermore, the manufacturing restrictions of PBF-LB/M are considered in the methodology, thus making the common extra step of adaptation of a component design with regard to this aspect [149] obsolete. This is also connected to reduced personnel costs and reduced lead time. By simultaneously considering various aspects, the methodology ensures a high quality of results that would be difficult to achieve without its application.

Furthermore, the methodology allows for an easily-documentable component development process that has high reproducibility. This is of importance as component design is relevant for certification in the aerospace industry [150]. In this way, the lead time to achieve a certified component can be reduced and personnel costs connected to the certification and documentation process can be saved.

The previously mentioned aspects show the great potential provided by the biomimetic design methodology with regard to the four important factors in project management and product development.

4.13.5 Limitations

One limitation of this work is given by the focus of the methodology on truss-like topology optimization designs as inputs. Even truss-like structures are a common output of topology optimization, wall-like structures might be part of the result especially for design spaces that are rather restricted. Topology optimization results containing wall-like structures are not suitable for the design methodology presented in this thesis and, therefore, should not be used as an input for it. Another limitation is given by the abstraction methodology as it is sensitive to the chosen abstraction parameters. Even though this is a common limitation in skeletonization and 3D abstraction methods, it is a noteworthy limitation of this work considering its application in the overall biomimetic design methodology.

With regard to beam design and beam optimization, the work is limited to two conventional and two biomimetic parametric beams. A greater selection of parametric beams that consider the manufacturing restrictions of PBF-LB/M would increase the solutions space of the design methodology, adding potential for the improvement of the lightweight character of the generated biomimetic component designs. Another limiting factor of this work is given by the node design methodology. According to the FEA results, the adapted nodes are responsible for the majority of critical stresses, as most critical regions are located inside the nodes or close to the nodes.

4.13.6 Future prospects

Reflecting on the previously presented limitations, future research should include additional parametric beam designs into the design methodology to improve its potential for the generation of lightweight component designs. Beams that might be considered are conventional beams such as I-beams, and rectangular hollow beams. Furthermore, biomimetic beams could be considered such as beams that are based on grass-stalks or bamboo. Additionally, beam designs that were generated by topology optimization and which were then further developed into parametric beam designs could be considered. Beam designs that enable direct removal of powder from the beam, eliminating the need to extract it through the nodes, are of significant interest. If powder extraction through the nodes were not necessary, there would be greater flexibility to improve the node design methodology.

Furthermore, the abstraction methodology should be improved to make it more robust and less sensitive to preselected abstraction parameters. Moreover, future research should improve the node design methodology. The majority of regions that were critical with regard to mechanical stresses in the three biomimetic component designs were located close to the nodes or were part of the nodes. The improved node design methodology should focus on an improved transition between node designs and beam designs as well as achieve a lattice structure that prevents peak stresses more efficiently. Improvement of the node design submethodology has a high relevance for improvement of the overall biomimetic design methodology. It can be assumed that a new methodology focusing on these aspects can solve current shortcomings of the methodology such that a mass reduction by biomimetic design is possible, while structural integrity can be ensured.

Finally, future work should consider the experimental validation of the biomimetic design methodology. Research efforts should focus on designing multiple biomimetic components using this methodology and performing their additive manufacturing, while evaluating geometric accuracy and powder removability after manufacture. Furthermore, experimental mechanical testing of multiple additively manufactured components should be performed, and the experimental results should be compared with those from FEA.

4.13.7 Resolution of the research questions

Reflecting on the primary research question, it can be concluded that:

> *Through the development of detailed submethodologies that together form a biomimetic design methodology, which successfully generated three biomimetic component designs from topology optimization designs, the primary research question of this thesis, was answered.*

In addressing the secondary research question, the analysis revealed that:

> *Through FEA, it was found that biomimetic designs did not maintain structural integrity in critical stress regions; however, they achieved a 12.5 % - 30.3 % reduction in weight compared to the topology optimization designs, highlighting a trade-off in lightweight design that addresses the secondary research question.*

A more general yet detailed methodology for the design of biomimetic additively manufactured components was presented in this thesis. It enables engineers with relatively little experience in the field, to create biomimetic component designs, thus considerably reducing barriers that previously existed. This will lead to a wider adoption of biomimetic additively manufactured components contributing to more sustainable and more efficient mechanical components in industries such as aerospace and automotive.

5. Conclusion and outlook

Previous research suggested that incorporating biomimetic beam-like structures into topology optimization designs can improve the component's lightweight characteristics. From this, two research questions were developed that are the basis of this thesis. The primary research question was how to develop a biomimetic design methodology to redesign mechanical components generated from density-based topology optimization by substituting cylindrical beams with biomimetic beams, while considering manufacturability in PBF-LB/M. The secondary research question was how biomimetic designs generated using such a methodology compare to the original topology optimization designs in terms of their lightweight character.

A five-step biomimetic design methodology for structural components manufactured in the PBF-LB/M process was developed, implemented and assessed. Its input is a density-based topology optimization result. Three common mechanical problem formulations were used to generate exemplary topology optimization results. In the first step of the biomimetic design methodology, an abstraction of the topology optimization is generated. A skeletonization approach is used to generate an interim abstraction by lines and points. In the next step, an abstraction by cylindrical beams and spherical nodes is generated. In the second step of the five-step methodology, the abstraction is subjected to the original mechanical boundary conditions of the topology optimization, an FEA is performed, and the internal forces and moments in each of the cylindrical beams are evaluated. Then load cases for parameter optimization of beams are derived. In step three of the biomimetic design methodology, two novel and two true biomimetic parametric beam designs based on the pseudo-stem of the banana plant were developed. These two biomimetic as well as the two conventional beam designs are considered for parameter optimization for each of the previously developed load cases. Numerous parameter optimizations are performed and the most lightweight optimized beam is selected for each of the beams of the abstraction, respectively. Buckling analysis is performed in this step for each selected beam under compression, and re-dimensioning is performed if necessary. In the fourth step of the methodology, node designs for the biomimetic component are generated. Solid volume node designs are generated and then a lattice infill structure is incorporated to increase the lightweight character of the nodes and to ensure powder removability. The nodes combine the beams to obtain a complete biomimetic component design. In step five of the methodology, validation of the biomimetic component design by FEA is performed and the mass is evaluated and compared to the mass of the input topology optimization design. It is noted that in all steps of the five-step design methodology the manufacturing restrictions of the PBF-LB/M process were considered.

The developed design methodology was applied to three initially generated topology optimization designs. Interim design results and complete biomimetic component designs could be generated successfully. FEA yielded that there are critical regions of high stress in all three of the biomimetic component designs. However, the locations of critical stresses have been associated with specific regions within the biomimetic designs. Mass evaluation showed that the biomimetic component designs are 12.5 % - 30.3 % lighter compared to their topology optimization counterparts. Achieving such a significant mass reduction is a challenging task, as topology optimization results approximate a mathematical optimum.

T. Röver, *Methodology for Biomimetic Design of Additively Manufactured Structural Components*, Light Engineering für die Praxis, https://doi.org/10.1007/978-3-032-27405-2_5

Comparing the presented methodology with other methodologies for redesign of topology optimization designs, it stands out as it includes the evaluation of internal forces and moments and the subsequent lightweight optimization for each beam of the design. Furthermore, the detailed design methodology complements previously existing biomimetic design methods that are rather coarse grained.

All four papers of this thesis, as well as pertinent research data, have been published as open access, making them highly accessible, allowing research and industry to directly build on this work. The methodology could be applied across various lightweight industries for components such as aerospace brackets and automotive vehicle structures. The author sees great potential offered by the methodology with regard to the four key factors: time, cost, quality, and sustainability. Direct and indirect weight savings reduce production and operating costs while decreasing noise and CO_2 emissions during operation improving customer comfort and environmental sustainability. The optimized digital value chain integrates a detailed design methodology with partial automation, design constraints of the PBF-LB/M process, and the conversion of topology optimization results to easily-editable component designs. Furthermore, the methodology is easily documented and has high reproducibility, which reduces the barriers for certification processes. The previously mentioned aspects reduce lead time and personnel costs, while ensuring high quality component designs. The presented design methodology could be adapted for the design of polymer or metal components that are produced by AM technologies other than PBF-LB/M. Moreover, the author sees a considerable potential for using the presented concepts and methodologies in architecture and civil engineering.

Future research should focus on the improvement of the node design methodology as the majority of regions that were critical with regard to mechanical stresses in the three biomimetic component designs were located close to the nodes or were part of them. Improvement of the node design submethodology is highly relevant for improvement of the overall biomimetic design methodology. Additionally, future research should assess components developed with the biomimetic design methodology experimentally after AM.

In conclusion, the thesis addressed the two formulated research questions successfully. A novel biomimetic design methodology for the development of lightweight AM components was developed and assessed. It is well accessible and engineers with little experience are enabled to generate highly complex biomimetic designs from topology optimization results to achieve additional weight savings of up to 30 %. Thus, it contributes to realizing the great potential of biomimetic design in lightweight industries such as aerospace and automotive.

References

[1] D.I. Cleland, Why project management?, Business Horizons 7 (1964) 81–88. https://doi.org/10.1016/0007-6813(64)90053-9.

[2] P. Weaver, THE ORIGINS OF MODERN PROJECT MANAGEMENT, in: Marriott Pinnacle Downtown, Vancouver., 2007.

[3] J.B. Ebbesen, A.J. Hope, Re-imagining the Iron Triangle: Embedding Sustainability into Project Constraints, PM World Journal 2 (2013). https://pmworldlibrary.net/wp-content/uploads/2013/03/pmwj8-mar2013-ebbesen-hope-reimagining-iron-triangle-FeaturedPaper1.pdf.

[4] The European Parliament and the Council of the European Union, REGULATION (EU) 2021/1119 of the European Parliament and the Council of 30 June 2021: establishing the framework for achieving climate neutrality and amending Regulations (EC) No 401/2009 and (EU) 2018/1999 ('European Climate Law'), Official Journal of the European Union (2021). https://eur-lex.europa.eu/legal-content/EN/TXT/?uri=CELEX:32021R1119.

[5] M.A. Carruth, J.M. Allwood, R.L. Milford, Reducing CO2 Emissions through Lightweight Design and Manufacturing, in: Belfast, (United Kingdom), 2011: pp. 1632–1637. https://doi.org/10.1063/1.3589750.

[6] R. Huang, M. Riddle, D. Graziano, J. Warren, S. Das, S. Nimbalkar, J. Cresko, E. Masanet, Energy and emissions saving potential of additive manufacturing: the case of lightweight aircraft components, Journal of Cleaner Production 135 (2016) 1559–1570. https://doi.org/10.1016/j.jclepro.2015.04.109.

[7] Md.F. Hossain, Global Sustainability, Springer International Publishing, Cham, 2023. https://doi.org/10.1007/978-3-031-34575-3.

[8] R.J. Yang, A.I. Chahande, Automotive applications of topology optimization, Structural Optimization 9 (1995) 245–249. https://doi.org/10.1007/BF01743977.

[9] J.-H. Zhu, W.-H. Zhang, L. Xia, Topology Optimization in Aircraft and Aerospace Structures Design, Archives of Computational Methods in Engineering 23 (2016) 595–622. https://doi.org/10.1007/s11831-015-9151-2.

[10] C. Emmelmann, M. Petersen, J. Kranz, E. Wycisk, Bionic lightweight design by laser additive manufacturing (LAM) for aircraft industry, in: P. Ambs, D. Curticapean, C. Emmelmann, W. Knapp, Z.T. Kuznicki, P.P. Meyrueis (Eds.), SPIE Eco-Photonics 2011: Sustainable Design, Manufacturing, and Engineering Workforce Education for a Green Future, SPIE, 2011: p. 80650. https://doi.org/10.1117/12.898525.

[11] M. Gralow, F. Weigand, D. Herzog, T. Wischeropp, C. Emmelmann, Biomimetic design and laser additive manufacturing—A perfect symbiosis?, Journal of Laser Applications 32 (2020) 021201. https://doi.org/10.2351/1.5131642.

[12] EOS GmbH, Certified for Universal Success: Additive Manufacturing of Satellite Components, (2018). https://www.eos.info/de/var/assets/01_parts-and-applications/case_studies_applications_parts/_case_studies_pdf/en_cases/cs_m_aerospace_ruag_en.pdf (accessed December 16, 2024).

[13] Inovar Communications Ltd, Airbus Defence and Space develops aluminium bracket for new Eurostar E3000 satellite platforms, (2015). https://www.metal-am.com/airbus-defence-and-space-develops-aluminium-bracket-for-new-eurostar-e3000-satellite-platforms/ (accessed December 17, 2024).

T. Röver, *Methodology for Biomimetic Design of Additively Manufactured Structural Components*, Light Engineering für die Praxis, https://doi.org/10.1007/978-3-032-27405-2

[14] 3D Systems Inc., Thales Alenia Space and 3D Systems Prove Viability of Direct Metal Printing, (n.d.). https://www.3dsystems.com/learning-center/case-studies/tipping-point (accessed December 17, 2024).

[15] Universität Paderborn, AM for satellites: Reaction Wheel Bracket, (n.d.). https://dmrc.uni-paderborn.de/en/innovation/produktentwicklung-und-anwendung/06-reaction-wheel-bracket (accessed December 17, 2024).

[16] J. Zhu, H. Zhou, C. Wang, L. Zhou, S. Yuan, W. Zhang, A review of topology optimization for additive manufacturing: Status and challenges, Chinese Journal of Aeronautics 34 (2021) 91–110. https://doi.org/10.1016/j.cja.2020.09.020.

[17] EOS North America, EOS Additive Manufacturing – A New Way Forward, (n.d.). https://additivesummit.com/wp-content/uploads/2019/08/2-EOS-Presentation.pdf (accessed December 16, 2024).

[18] European Space Agency, Metal 3D-printed Ariane launcher bracket, (2019). https://www.esa.int/ESA_Multimedia/Images/2019/06/Metal_3D-printed_Ariane_launcher_bracket (accessed December 16, 2024).

[19] APWORKS GmbH, Journey of the Light Rider, (n.d.). https://www.apworks.de/blog-lightrider (accessed December 16, 2024).

[20] Czinger Vehicles Inc., Czinger - revolutionary technology, (n.d.). https://www.czinger.com/revolutionary-technology (accessed December 17, 2024).

[21] M. Hunar, L. Jancar, D. Krzikalla, D. Kaprinay, D. Srnicek, Comprehensive View on Racing Car Upright Design and Manufacturing, Symmetry 12 (2020) 1020. https://doi.org/10.3390/sym12061020.

[22] Canyon Bicycles GmbH, CANYON showcases visionary sustainable bike frame developed with bike magazin, (2022). https://media-centre.canyon.com/en-INT/220598-canyon-showcases-visionary-sustainable-bike-frame-developed-with-bike-magazin (accessed December 17, 2024).

[23] T. Röver, R.J. Lau, F. Lange, A. Struve, C. Fuchs, K. Bartsch, A. Seibel, C. Emmelmann, Methodology for Integrating Biomimetic Beams in Abstracted Topology Optimization Results, in: Proceedings of the ASME 2022 International Mechanical Engineering Congress and Exposition. Volume 4: Biomedical and Biotechnology; Design, Systems, and Complexity, Columbus, Ohio, USA, 2022. https://doi.org/10.1115/IMECE2022-94299.

[24] T. Röver, M. Bader, K. Asami, C. Emmelmann, I. Kelbassa, Development and assessment of a methodology for abstraction of topology optimization results to enable the substitution of optimized beams, Journal of Laser Applications 35 (2023). https://doi.org/10.2351/7.0001185.

[25] T. Röver, C. Fuchs, K. Asami, C. Emmelmann, Dimensioning of Biomimetic Beams under Bending for Additively Manufactured Structural Components, Biomimetics 9 (2024) 214. https://doi.org/10.3390/biomimetics9040214.

[26] T. Röver, T. Sievers, K. Asami, C. Emmelmann, Novel Lightweight Design of Additively Manufactured Structural Components through Biomimetics, Journal of the Japan Society of Powder and Powder Metallurgy 72 (2025) S1241–S1248. https://doi.org/10.2497/jjspm.16C-T11-10.

[27] DIN EN ISO/ASTM 52900:2022-03 Additive manufacturing - General principles - Fundamentals and vocabulary (ISO/ASTM 52900:2021); German version EN ISO/ASTM 52900:2021, (2022). https://doi.org/10.31030/3290011.

[28] F. Huber, M. Rasch, M. Schmidt, Laser Powder Bed Fusion (PBF-LB/M) Process Strategies for In-Situ Alloy Formation with High-Melting Elements, Metals 11 (2021) 336. https://doi.org/10.3390/met11020336.

[29] C. Brecher, ed., Integrative Produktionstechnik für Hochlohnländer, Springer Berlin Heidelberg, Berlin, Heidelberg, 2011. https://doi.org/10.1007/978-3-642-20693-1.

[30] D. Herzog, V. Seyda, E. Wycisk, C. Emmelmann, Additive manufacturing of metals, Acta Materialia 117 (2016) 371–392. https://doi.org/10.1016/j.actamat.2016.07.019.

[31] I. Gibson, D. Rosen, B. Stucker, M. Khorasani, Additive Manufacturing Technologies, Springer International Publishing, Cham, 2021. https://doi.org/10.1007/978-3-030-56127-7.

[32] M. Shellabear, O. Nyrhilä, DMLS – DEVELOPMENT HISTORY AND STATE OF THE ART, in: Erlangen, Germany, 2004. https://www.i3dmfg.com/wp-content/uploads/2015/07/History-of-DMLS.pdf.

[33] M. Munsch, M. Schmidt-Lehr, E. Wycisk, AMPOWER REPORT 2024 - Additive Manufacturing Market Report, AMPOWER GmbH & Co. KG, 2024.

[34] E. Celik, Additive Manufacturing: Science and Technology, De Gruyter, Berlin; Boston, 2020. https://doi.org/10.1515/9781501518782.

[35] Wohlers Associates, ed., Wohlers report 2019: 3D printing and additive manufacturing state of the industry, Wohlers Associates, Fort Collins (Colorado), 2019.

[36] M. Leary, Design for additive manufacturing, Elsevier, Amsterdam Kidlington, Oxford Cambridge, MA, 2020.

[37] J. Kranz, D. Herzog, C. Emmelmann, Design guidelines for laser additive manufacturing of lightweight structures in TiAl6V4, Journal of Laser Applications 27 (2015) 14001. https://doi.org/10.2351/1.4885235.

[38] D. Herzog, K. Asami, C. Scholl, C. Ohle, C. Emmelmann, A. Sharma, N. Markovic, A. Harris, Design guidelines for laser powder bed fusion in Inconel 718, Journal of Laser Applications 34 (2022) 012015. https://doi.org/10.2351/7.0000508.

[39] DIN EN ISO/ASTM 52910:2020-02 Additive manufacturing - Design - Requirements, guidelines and recommendations (ISO/ASTM 52910:2018); German version EN ISO/ASTM 52910:2019, (2020). https://doi.org/10.31030/3111254.

[40] DIN EN ISO/ASTM 52911-1:2020-05 Additive manufacturing - Design - Part 1: Laser-based powder bed fusion of metals (ISO/ASTM 52911-1:2019); German version EN ISO/ASTM 52911-1:2019, (2020). https://doi.org/10.31030/3060962.

[41] E. Atzeni, A. Salmi, Study on unsupported overhangs of AlSi10Mg parts processed by Direct Metal Laser Sintering (DMLS), Journal of Manufacturing Processes 20 (2015) 500–506. https://doi.org/10.1016/j.jmapro.2015.04.004.

[42] Q. Han, H. Gu, S. Soe, R. Setchi, F. Lacan, J. Hill, Manufacturability of AlSi10Mg overhang structures fabricated by laser powder bed fusion, Materials & Design 160 (2018) 1080–1095. https://doi.org/10.1016/j.matdes.2018.10.043.

[43] A. Seharing, A.H. Azman, S. Abdullah, A review on integration of lightweight gradient lattice structures in additive manufacturing parts, Advances in Mechanical Engineering 12 (2020). https://doi.org/10.1177/1687814020916951.

[44] J. Wu, O. Sigmund, J.P. Groen, Topology optimization of multi-scale structures: a review, Structural and Multidisciplinary Optimization 63 (2021) 1455–1480. https://doi.org/10.1007/s00158-021-02881-8.

[45] L.-Y. Chen, S.-X. Liang, Y. Liu, L.-C. Zhang, Additive manufacturing of metallic lattice structures: Unconstrained design, accurate fabrication, fascinated performances, and challenges, Materials Science and Engineering: R: Reports 146 (2021) 100648. https://doi.org/10.1016/j.mser.2021.100648.

[46] J. Feng, J. Fu, X. Yao, Y. He, Triply periodic minimal surface (TPMS) porous structures: from multi-scale design, precise additive manufacturing to multidisciplinary

applications, International Journal of Extreme Manufacturing 4 (2022) 022001. https://doi.org/10.1088/2631-7990/ac5be6.

[47] M. Zhang, J. Shang, S. Guo, B. Hur, X. Yue, Numerical Investigation of Effective Thermal Conductivity of Strut-Based Cellular Structures Designed by Spatial Voronoi Tessellation, Materials 14 (2020) 138. https://doi.org/10.3390/ma14010138.

[48] W.R. Spillers, K.M. MacBain, Structural optimization, Springer, Dordrecht; New York, 2009.

[49] L.A. Schmit, STRUCTURAL DESIGN BY SYSTEMATIC SYNTHESIS, in: Pittsburgh, Pennsylvania, USA, 1960. https://www.vrand.com/wp-content/uploads/2017/06/Struc_Design_Sys_Synthesis.pdf (accessed August 2, 2024).

[50] M.P. Bendsøe, O. Sigmund, Topology Optimization: Theory, Methods, and Applications, Second Edition, Corrected Printing, Springer, Berlin; Heidelberg, 2004. https://doi.org/10.1007/978-3-662-05086-6.

[51] U. Kirsch, Structural optimization: fundamentals and applications, Springer-Verlag, Berlin, 1993.

[52] O. Sigmund, K. Maute, Topology optimization approaches: A comparative review, Structural and Multidisciplinary Optimization 48 (2013) 1031–1055. https://doi.org/10.1007/s00158-013-0978-6.

[53] M.P. Bendsøe, N. Kikuchi, Generating optimal topologies in structural design using a homogenization method, Computer Methods in Applied Mechanics and Engineering 71 (1988) 197–224. https://doi.org/10.1016/0045-7825(88)90086-2.

[54] M.P. Bendsøe, Optimal shape design as a material distribution problem, Structural Optimization 1 (1989) 193–202. https://doi.org/10.1007/BF01650949.

[55] COMSOL, COMSOL Multiphysics - Optimization Module User's Guide: Version: 6.0, (2021). https://doc.comsol.com/6.0/doc/com.comsol.help.opt/OptimizationModuleUsersGuide.pdf.

[56] Dassault Systèmes, SIMP Method for Topology Optimization - 2023 - SOLIDWORKS Help, (n.d.). https://help.solidworks.com/2023/english/SolidWorks/cworks/c_simp_method_topology.htm#vdl1527111485501 (accessed June 19, 2024).

[57] Siemens Product Lifecycle Management Software Inc, Simcenter Nastran Topology Optimization, (n.d.). https://feacomp.com/wp-content/uploads/2023/03/6.-Simcenter-Nastran-Topology-Optimization-Fact-Sheet.pdf (accessed June 19, 2024).

[58] J. Kranz, Methodik und Richtlinien für die Konstruktion von laseradditiv gefertigten Leichtbaustrukturen, Springer-Verlag GmbH, Berlin, 2017. https://doi.org/10.1007/978-3-662-55339-8.

[59] B.S. Lazarov, O. Sigmund, Filters in topology optimization based on Helmholtz-type differential equations, International Journal for Numerical Methods in Engineering 86 (2011) 765–781. https://doi.org/10.1002/nme.3072.

[60] J.D. Deaton, R.V. Grandhi, A survey of structural and multidisciplinary continuum topology optimization: post 2000, Structural and Multidisciplinary Optimization 49 (2014) 1–38. https://doi.org/10.1007/s00158-013-0956-z.

[61] T. Röver, Development of a Methodology for Substitution of Stress Fields from Topology Optimization by Parametric Bionic Structures, Student thesis - Master thesis, Hamburg University of Technology, 2019.

[62] D. Walton, H. Moztarzadeh, Design and Development of an Additive Manufactured Component by Topology Optimisation, Procedia CIRP 60 (2017) 205–210. https://doi.org/10.1016/j.procir.2017.03.027.

[63] D. Herzog, T. Röver, S. Abdolov, F. Becker, C. Gentner, Optimization and design for additive manufacturing of a fuel cell end plate, Journal of Laser Applications 34 (2022) 042027. https://doi.org/10.2351/7.0000789.

[64] T. Dbouk, A review about the engineering design of optimal heat transfer systems using topology optimization, Applied Thermal Engineering 112 (2017) 841–854. https://doi.org/10.1016/j.applthermaleng.2016.10.134.

[65] T. Röver, S. Roth, D. Herzog, T. Hoffmann, L. Baetcke, Development of a Hydrogen Metal Hydride Storage Produced by Additive Manufacturing, in: European Fuel Cell Forum 2023: Low-Temp. Fuel Cells, Electrolysers & H2 Processing, Lucerne, Switzerland, 2023. https://doi.org/10.15480/882.9410.

[66] T. Röver, S. Roth, D. Herzog, T. Hoffmann, Poster: Development of a Hydrogen Metal Hydride Storage Produced by Additive Manufacturing, (2023). https://doi.org/10.15480/882.9411.

[67] T. Röver, S. Roth, T. Hoffmann, L. Baetcke, D. Herzog, Supplementary material to extended abstract with title: Development of a Hydrogen Metal Hydride Storage Produced by Additive Manufacturing, (2023). https://doi.org/10.15480/336.5136.

[68] F. Lange, C. Hein, G. Li, C. Emmelmann, Numerical optimization of active heat sinks considering restrictions of selective laser melting, in: Proceedings of the 2018 COMSOL Conference in Lausanne, 2018. https://www.comsol.com/paper/download/569351/lange_paper.pdf.

[69] A. Fawaz, Y. Hua, S. Le Corre, Y. Fan, L. Luo, Topology optimization of heat exchangers: A review, Energy 252 (2022) 124053. https://doi.org/10.1016/j.energy.2022.124053.

[70] N. Wu, S. Li, B. Zhang, C. Wang, B. Chen, Q. Han, J. Wang, The advances of topology optimization techniques in orthopedic implants: A review, Medical & Biological Engineering & Computing 59 (2021) 1673–1689. https://doi.org/10.1007/s11517-021-02361-7.

[71] F. Lucchini, R. Torchio, V. Cirimele, P. Alotto, P. Bettini, Topology Optimization for Electromagnetics: A Survey, IEEE Access 10 (2022) 98593–98611. https://doi.org/10.1109/ACCESS.2022.3206368.

[72] J. Alexandersen, C.S. Andreasen, A Review of Topology Optimisation for Fluid-Based Problems, Fluids 5 (2020) 29. https://doi.org/10.3390/fluids5010029.

[73] D. Dinkler, U. Kowalsky, Introduction to Finite Element Methods, Springer Fachmedien Wiesbaden, Wiesbaden, 2024. https://doi.org/10.1007/978-3-658-42742-9.

[74] J. Bielak, The Finite Element Method: A Primer, Springer International Publishing, Cham, 2024. https://doi.org/10.1007/978-3-031-56369-0.

[75] F. Wang, B.S. Lazarov, O. Sigmund, On projection methods, convergence and robust formulations in topology optimization, Structural and Multidisciplinary Optimization 43 (2011) 767–784. https://doi.org/10.1007/s00158-010-0602-y.

[76] J.-C. Cuillière, V. François, A. Nana, Automatic construction of structural CAD models from 3D topology optimization, Computer-Aided Design and Applications 15 (2018) 107–121. https://doi.org/10.1080/16864360.2017.1353726.

[77] H. Blum, A transformation for extracting new descriptors of shape, in: Models for the Perception of Speech and Visual Form, MIT Press, Cambridge, Massachusetts, 1967: pp. 362–380.

[78] P.K. Saha, G. Borgefors, G. Sanniti di Baja, eds., Skeletonization: theory, methods and applications, Academic Press, an imprint of Elsevier, London; San Diego; Cambridge, Massachusetts, 2017.

[79] N.D. Cornea, D. Silver, P. Min, Curve-skeleton properties, applications, and algorithms, IEEE Transactions on Visualization and Computer Graphics 13 (2007) 530–48. https://doi.org/10.1109/TVCG.2007.1002.

[80] A. Tagliasacchi, T. Delame, M. Spagnuolo, N. Amenta, A. Telea, 3D Skeletons: A State-of-the-Art Report, Computer Graphics Forum 35 (2016) 573–597. https://doi.org/10.1111/cgf.12865.

[81] D.G. Morgenthaler, Three-Dimensional Simple Points: Serial Erosion, Parallel Thinning and Skeletonization, Computer Vision Lab., University of Maryland, Maryland, 1981.

[82] F. Lazarus, A. Verroust, Level set diagrams of polyhedral objects, in: Proceedings of the Fifth ACM Symposium on Solid Modeling and Applications, ACM, Ann Arbor, Michigan, USA, 1999: pp. 130–140. https://doi.org/10.1145/304012.304025.

[83] N. Ahuja, J.-H. Chuang, Shape representation using a generalized potential field model, IEEE Transactions on Pattern Analysis and Machine Intelligence 19 (1997) 169–176. https://doi.org/10.1109/34.574801.

[84] N.D. Cornea, D. Silver, X. Yuan, R. Balasubramanian, Computing hierarchical curve-skeletons of 3D objects, The Visual Computer 21 (2005) 945–955. https://doi.org/10.1007/s00371-005-0308-0.

[85] A. Nana, J.-C. Cuillière, V. Francois, Automatic reconstruction of beam structures from 3D topology optimization results, Computers & Structures 189 (2017) 62–82. https://doi.org/10.1016/j.compstruc.2017.04.018.

[86] Synera GmbH, Synera Platform, (2024). https://www.synera.io/platform (accessed September 20, 2024).

[87] Synera | Connected Engineering, 09 - Topology Reconstruction (Part 2) - Quad Skeleton Process, (2023). https://www.youtube.com/watch?v=bFx6ZtooJho&list=PL_6y2PVnbDrl7xqWlU7sXDXayOn3Ry44R&index=10 (accessed September 20, 2024).

[88] J.M. Harkness, In Appreciation - A Lifetime of Connections: Otto Herbert Schmitt, 1913 - 1998, Physics in Perspective (PIP) 4 (2002) 456–490. https://doi.org/10.1007/s000160200005.

[89] J.F.V. Vincent, O.A. Bogatyreva, N.R. Bogatyrev, A. Bowyer, A.-K. Pahl, Biomimetics: its practice and theory, Journal of The Royal Society Interface 3 (2006) 471–482. https://doi.org/10.1098/rsif.2006.0127.

[90] BIONICS SYMPOSIUM -Living prototypes - The key to new technology, Wright Air Development Devision, 1960.

[91] A. Iouguina, J.W. Dawson, B. Hallgrimsson, G. Smart, Biologically informed disciplines: a comparative analysis of bionics, biomimetics, biomimicry, and bio-inspiration among others, International Journal of Design & Nature and Ecodynamics 9 (2014) 197–205. https://doi.org/10.2495/DNE-V9-N3-197-205.

[92] VDI-Gesellschaft Technologies of Life Sciences, VDI 6220 Blatt 1:2021-09 - Biomimetics - Fundamentals, conception, and strategy, (2021). https://www.dinmedia.de/de/technische-regel/vdi-6220-blatt-1/335605718 (accessed July 3, 2024).

[93] VDI-Gesellschaft Technologies of Life Sciences, VDI 6224 Blatt 3:2017-09 - Biomimetics - Integrated product development process for biomimetic optimisation, (2017). https://www.dinmedia.de/de/technische-regel/vdi-6224-blatt-3/269122178 (accessed July 11, 2024).

[94] C. Hamm, S. Möller, ELiSE—An Integrated, Holistic Bionic Approach to Develop Optimized Lightweight Solutions for Engineering, Architecture and Design, in: C.

Hamm (Ed.), Evolution of Lightweight Structures, Springer Netherlands, Dordrecht, 2015: pp. 183–194. https://doi.org/10.1007/978-94-017-9398-8_11.

[95] A. Du Plessis, C. Broeckhoven, I. Yadroitsava, I. Yadroitsev, C.H. Hands, R. Kunju, D. Bhate, Beautiful and Functional: A Review of Biomimetic Design in Additive Manufacturing, Additive Manufacturing 27 (2019) 408–427. https://doi.org/10.1016/j.addma.2019.03.033.

[96] T.M. Wischeropp, H. Hoch, F. Beckmann, C. Emmelmann, Opportunities for Braking Technology Due to Additive Manufacturing Through the Example of a Bugatti Brake Caliper, in: R. Mayer (Ed.), XXXVII. Internationales μ-Symposium 2018 Bremsen-Fachtagung, Springer Berlin Heidelberg, Berlin; Heidelberg, 2019: pp. 181–193. https://doi.org/10.1007/978-3-662-58024-0_12.

[97] C. Klein, Weltpremiere der Fertigungstechnik - Titan-Teil: Bugatti bringt Bremssattel aus dem 3D-Drucker, (2018). https://www.automobil-produktion.de/produktion/titan-teil-bugatti-bringt-bremssattel-aus-dem-3d-drucker-120.html.

[98] Pioneering bionic 3D printing, (2016). https://www.airbus.com/en/newsroom/news/2016-03-pioneering-bionic-3d-printing (accessed July 16, 2024).

[99] D. Gross, W. Hauger, J. Schröder, W.A. Wall, Technische Mechanik. 2: Elastostatik, 13., aktualisierte Auflage, Springer Vieweg, Berlin; Heidelberg, 2017. https://doi.org/10.1007/978-3-662-53679-7.

[100] L. Euler, Methodus inveniendi lineas curvas maximi minimive proprietate gaudentes, sive solutio problematis isoperimetrici lattissimo sensu accepti, Lausannæ & Genevæ, 1744. https://scholarlycommons.pacific.edu/euler-works/65.

[101] J.M. Gere, S.P. Timoshenko, Mechanics of materials, 4. ed, PWS PUBLISHING COMPANY, Boston, 1997.

[102] J. Ma, W. Chen, L. Zhao, D. Zhao, Elastic Buckling of Bionic Cylindrical Shells Based on Bamboo, Journal of Bionic Engineering 5 (2008) 231–238. https://doi.org/10.1016/S1672-6529(08)60029-3.

[103] G.N. Karam, L.J. Gibson, Elastic buckling of cylindrical shells with elastic cores—I. Analysis, International Journal of Solids and Structures 32 (1995) 1259–1283. https://doi.org/10.1016/0020-7683(94)00147-O.

[104] W. Nachtigall, A. Wisser, Bionics by examples, Springer Berlin Heidelberg, New York, NY, 2014.

[105] D. Pleasant, C. Gavin, G. Redden, J. Nagel, H. Zhang, Bioinspired Design of Material Architecture for Additive Manufacturing, Machines 11 (2023) 1081. https://doi.org/10.3390/machines11121081.

[106] D. Ashok, M.V.A.R. Bahubalendruni, Design and Characterization of 2.5D Nature-Inspired Infill Structures under Out-Plane Quasi-Static Loading Condition, Advances in Materials Science and Engineering 2023 (2023) 1–11. https://doi.org/10.1155/2023/8918937.

[107] W. Nachtigall, A. Wisser, Biologisches Design, Springer Berlin Heidelberg, 2005.

[108] VDI/VDE-Gesellschaft Mess- und Automatisierungstechnik (GMA), VDI/VDE 2206:2021-11 - Development of mechatronic and cyber-physical systems, (2021). https://www.dinmedia.de/de/technische-regel/vdi-vde-2206/342674320 (accessed July 23, 2024).

[109] R. Stephens, Beginning software engineering, Wrox, A Wiley Brand, Indianapolis, Indiana, 2015.

[110] COMSOL, COMSOL Multiphysics - Reference Manual - Version: 5.4, (2018). https://doc.comsol.com/5.4/doc/com.comsol.help.comsol/COMSOL_Reference-Manual.pdf.

[111] COMSOL, COMSOL Multiphysics - Reference Manual - Version: 6.0, (2021). https://doc.comsol.com/6.0/doc/com.comsol.help.comsol/COMSOL_Reference-Manual.pdf.

[112] J. Hicklin, C. Moler, P. Webb, R.F. Boisvert, B. Miller, R. Pozo, K. Remington, JAMA: A Java Matrix Package: The MathWorks and the National Institute of Standards and Technology (NIST), (2012). https://math.nist.gov/javanumerics/jama/ (accessed July 31, 2024).

[113] Autodesk Inc., Inventor 2024 Help | Inventor API User's Manual | Autodesk, (n.d.). https://help.autodesk.com/view/INVNTOR/2024/ENU/?guid=UserManualIndex (accessed April 19, 2024).

[114] Autodesk Inc., AUTODESK Netfabb 2023, (2022). https://help.autodesk.com/view/NETF/2023/ENU/ (accessed July 31, 2024).

[115] nTopology Inc., Next-Gen Engineering Design Software: nTop (formerly nTopology), nTop (n.d.). https://www.ntop.com/ (accessed April 19, 2024).

[116] One Click Metal GmbH, BOLDSERIES - Technical Specifications - MPRINT, (n.d.). https://oneclickmetal.com/boldseries-technical-specifications/ (accessed July 31, 2024).

[117] N. Limbasiya, A. Jain, H. Soni, V. Wankhede, G. Krolczyk, P. Sahlot, A comprehensive review on the effect of process parameters and post-process treatments on microstructure and mechanical properties of selective laser melting of AlSi10Mg, Journal of Materials Research and Technology 21 (2022) 1141–1176. https://doi.org/10.1016/j.jmrt.2022.09.092.

[118] F. Trevisan, F. Calignano, M. Lorusso, J. Pakkanen, A. Aversa, E. Ambrosio, M. Lombardi, P. Fino, D. Manfredi, On the Selective Laser Melting (SLM) of the AlSi10Mg Alloy: Process, Microstructure, and Mechanical Properties, Materials 10 (2017) 76. https://doi.org/10.3390/ma10010076.

[119] SLM Solutions Group AG, Material Data Sheet Al-Alloy AlSi10Mg, (2022). https://www.slm-solutions.com/fileadmin/Content/Powder/MDS/MDS_Al-Alloy_AlSi10Mg_0520_EN.pdf (accessed August 24, 2022).

[120] E. Sert, E. Schuch, A. Öchsner, L. Hitzler, E. Werner, M. Merkel, Tensile strength performance with determination of the Poisson's ratio of additively manufactured AlSi10Mg samples, Materialwissenschaft Und Werkstofftechnik 50 (2019) 539–545. https://doi.org/10.1002/mawe.201800233.

[121] T. Röver, M. Bader, M.K. Asami, C. Emmelmann, I. Kelbassa, Supplementary material to article with title: Development and Assessment of a Methodology for Abstraction of Topology Optimization Results to Enable the Substitution of Optimized Beams, (2023). https://doi.org/10.15480/882.8168.

[122] T. Röver, C. Fuchs, M.K. Asami, C. Emmelmann, Parameter Optimization Files (COMSOL Multiphysics 6.0) for Dimensioning of Biomimetic Beams under Bending, (2024). https://doi.org/10.15480/882.9113.

[123] T. Röver, T. Sievers, M.K. Asami, C. Emmelmann, Design and Simulation Files for a Novel Lightweight Design of Additively Manufactured Structural Components through Biomimetics, (2025). https://doi.org/10.15480/882.13772.

[124] C. Fuchs, Optimierung bionischer Strukturen unter Berücksichtigung von Restriktionen additiver Fertigungstechnologien, Student thesis - Project thesis, Hamburg University of Technology, 2019.

[125] M. Bader, Überprüfung und Weiterentwicklung einer Abstraktionsmethodik zur Einbindung von bionischen Strukturen im Design von Bauteilen in der additiven Fertigung, Student thesis - Project thesis, Hamburg University of Technology, 2023.

[126] T. Sievers, Entwicklung und Bewertung einer Methodik zur Integration von bionischen Balken in abstrahierte Topologieoptimierungsergebnisse, Student thesis - Bachelor thesis, Hamburg University of Technology, 2024.

[127] G. Fiuk, M.W. Mrzygłód, Numerical benchmarks for topology optimization of structures with stress constraints, Bulletin of the Polish Academy of Sciences Technical Sciences (2021) 139317–139317. https://doi.org/10.24425/bpasts.2021.139317.

[128] O. Sigmund, A 99 line topology optimization code written in Matlab, Structural and Multidisciplinary Optimization 21 (2001) 120–127. https://doi.org/10.1007/s001580050176.

[129] SLM Solutions Group AG, SLM®800 Brochure - Selective Laser Melting Machine - Large Format Selective Laser Melting, (n.d.). https://www.slm-solutions.com/fileadmin/Content/Machines/SLM_R_800_Web.pdf (accessed August 1, 2024).

[130] COMSOL, COMSOL Multiphysics - Structural Mechanics Module - User's Guide - Version: 6.0, (2021). https://doc.comsol.com/6.0/doc/com.comsol.help.sme/StructuralMechanicsModuleUsersGuide.pdf (accessed September 24, 2024).

[131] foxtail_1, banana 1, (n.d.). https://openverse.org/image/4a31b903-ab5b-4485-9a97-3f1b4d65028c/.

[132] J.A. Nelder, R. Mead, A Simplex Method for Function Minimization, The Computer Journal 7 (1965) 308–313. https://doi.org/10.1093/comjnl/7.4.308.

[133] J.A. Nelder, R. Mead, Errata, The Computer Journal 8 (1965) 27–27. https://doi.org/10.1093/comjnl/8.1.27.

[134] W. Tao, M.C. Leu, Design of lattice structure for additive manufacturing, in: 2016 International Symposium on Flexible Automation (ISFA), IEEE, Cleveland, OH, USA, 2016: pp. 325–332. https://doi.org/10.1109/ISFA.2016.7790182.

[135] O. Rehme, C. Emmelmann, Rapid manufacturing of lattice structures with selective laser melting, in: F.G. Bachmann, W. Hoving, Y. Lu, K. Washio (Eds.), San Jose, CA, 2006: p. 61070K. https://doi.org/10.1117/12.645848.

[136] S.H. Asif, K. Hasan, N.R. Dhar, Topology optimization and 3D printing of a unibody quadcopter airframe, IOP Conference Series: Materials Science and Engineering 1305 (2024) 012021. https://doi.org/10.1088/1757-899X/1305/1/012021.

[137] S. Nvss, B. Esakki, L.-J. Yang, C. Udayagiri, K.S. Vepa, Design and Development of Unibody Quadcopter Structure Using Optimization and Additive Manufacturing Techniques, Designs 6 (2022) 8. https://doi.org/10.3390/designs6010008.

[138] J. Bright, R. Suryaprakash, S. Akash, A. Giridharan, Optimization of quadcopter frame using generative design and comparison with DJI F450 drone frame, IOP Conference Series: Materials Science and Engineering 1012 (2021) 012019. https://doi.org/10.1088/1757-899X/1012/1/012019.

[139] X. Yan, D. Bao, Y. Zhou, Y. Xie, T. Cui, Detail control strategies for topology optimization in architectural design and development, Frontiers of Architectural Research 11 (2022) 340–356. https://doi.org/10.1016/j.foar.2021.11.001.

[140] A. Jipa, M. Bernhard, M. Meibodi, B. Dillenburger, 3D-Printed Stay-in-Place Formwork for Topologically Optimized Concrete Slabs, (2016) 10 p. https://doi.org/10.3929/ETHZ-B-000237082.

[141] S. Bhoosan, Collaborative Design: Combining Computer-Aided Geometry Design and Building Information Modelling, Architectural Design 87 (2017) 82–89. https://doi.org/10.1002/ad.2177.

[142] H. Ohmori, Computational Morphogenesis: Its Current State and Possibility for the Future, International Journal of Space Structures 25 (2010) 75–82. https://doi.org/10.1260/0266-3511.25.2.75.

[143] S. Arslan, A.G. Sorguc, Similarities Between "structures In Nature" And "man-made Structures": Biomimesis In Architecture, 73 (2004) 10. https://doi.org/10.2495/DN040051.

[144] E. Torenbeek, Advanced Aircraft Design: Conceptual Design, Analysis and Optimization of Subsonic Civil Airplanes, 1st ed., Wiley, 2013. https://doi.org/10.1002/9781118568101.

[145] A.G. Seabridge, Design and development of aircraft systems, Third edition, Wiley, Hoboken, N.J., USA, 2020.

[146] U. Michel, Correlation of aircraft certification noise levels EPNL with controlling physical parameters, in: 19th AIAA/CEAS Aeroacoustics Conference, American Institute of Aeronautics and Astronautics, Berlin, Germany, 2013. https://doi.org/10.2514/6.2013-2014.

[147] J.-P. Rudolph, C. Emmelmann, Self-learning Calculation for Selective Laser Melting, Procedia CIRP 67 (2018) 185–190. https://doi.org/10.1016/j.procir.2017.12.197.

[148] D. Ochs, K.K. Wehnert, J. Hartmann, A. Schiffler, J. Schmitt, Sustainable Aspects of a Metal Printing Process Chain with Laser Powder Bed Fusion (LPBF), Procedia CIRP 98 (2021) 613–618. https://doi.org/10.1016/j.procir.2021.01.163.

[149] E. Dalpadulo, F. Gherardini, F. Pini, F. Leali, Integration of Topology Optimisation and Design Variants Selection for Additive Manufacturing-Based Systematic Product Redesign, Applied Sciences 10 (2020) 7841. https://doi.org/10.3390/app10217841.

[150] M. Hinsch, Guideline for EN 9100:2018: An Introduction to the European Aerospace and Defence Standard, Springer Berlin Heidelberg, Berlin, Heidelberg, 2020. https://doi.org/10.1007/978-3-662-61367-2.

Appendix A: Research articles

A.1 Methodology for integrating biomimetic beams in abstracted topology optimization results

The following pages feature a reprint of the conference article "Methodology for integrating biomimetic beams in abstracted topology optimization results," originally published in 2022. This reprint is provided under the terms of the CC BY 4.0 license.

T. Röver, *Methodology for Biomimetic Design of Additively Manufactured Structural Components*, Light Engineering für die Praxis, https://doi.org/10.1007/978-3-032-27405-2

Proceedings of the ASME 2022
International Mechanical Engineering Congress and Exposition
IMECE2022
October 30-November 3, 2022, Columbus, Ohio

IMECE2022-94299

METHODOLOGY FOR INTEGRATING BIOMIMETIC BEAMS IN ABSTRACTED TOPOLOGY OPTIMIZATION RESULTS

Tim Röver[1,2,*], Robert Johannes Lau[2], Fritz Lange[2], Arnd Struve[2], Cedrik Fuchs[1,2], Katharina Bartsch[1,2], Arthur Seibel[1,2], Claus Emmelmann[1]

[1]Hamburg University of Technology (TUHH), Institute of Laser and Systems Technologies (iLAS), Harburger Schloßstraße 28, 21079 Hamburg, Germany
[2]Fraunhofer Research Institution for Additive Manufacturing Technologies IAPT, Am Schleusengraben 14, 21029 Hamburg, Germany

ABSTRACT

This paper presents a five-step design methodology to generate designs of biomimetic structural components from topology optimization results. In step one, all material allocated by topology optimization is classified as either beam like structures or nodes to generate an auxiliary model consisting of preserved regions, cylindrical beams, and ball nodes, which is an abstraction of the original topology optimization result. In step two, the auxiliary model is exposed to the original boundary conditions in a finite element analysis. Then, internal forces, torsion, and bending moments in all beams of the auxiliary model are identified with respect to both of their ends. In step three, a database is used to find a suitable biomimetic beam for each previously analyzed beam in the auxiliary model. In step four, adapted nodes are designed to connect the biomimetic beams and preserved regions to generate an intermediate biomimetic component design. And in step five, a design iteration and a validation of the final design are performed. The design methodology allows for reproducible biomimetic component designs, a trackable and easily documentable component development process, and the possibility of automating the design process to ultimately save development costs when designing structural components.

Keywords: Topology optimization, FEM, beam structures, biomimetics, component design

1. INTRODUCTION

Topology optimization (TO) has become of great importance in the development of mechanical components in the last decades, especially for lightweight design in aerospace and automotive applications [1, 2]. Based on a design space and relevant boundary conditions, such as fixed surfaces, forces, and pressures, the topology optimization algorithm allocates material inside the design space such that the loads are well supported by the design. Using the minimum mass approach, only as much material as needed is allocated to support the loads without the material reaching a certain threshold, such as 80 % of the yield strength of the material at any point inside the component.

Previous works indicate that under certain load conditions, biomimetic beams may be beneficial in terms of structural integrity when being compared to cylindrical beams [3]. It is suggested that adding geometric complexity by integrating biomimetic beams into design concepts from topology optimizations may lead to designs with reduced mass. Considering components of up to 0.5 m length made from metals or plastics, production of these kinds of components can be done very effectively by additive manufacturing techniques, such as Powder Bed Fusion by Laser Beam of Metals (PBF-LB/M).

Although structural component designs incorporating biomimetic beams seem to have great potential, they are not widely used. The authors expect that this is due to absence of a corresponding design methodology. For this reason, a design methodology is proposed in this article in which topology optimization results are skeletonized and converted into new component designs by including biomimetic beams leading to designs of lower mass while sufficiently supporting the acting loads.

2. RELATED WORK

The methodology proposed in this article has parallels to and is partly built upon the methodologies from [4, 5]. Therefore, these methodologies are summarized in the following.

After a 3D topology optimization is performed, a smoothing algorithm is used to give the topology optimization a smoother surface. Then, a mesh contraction Laplacian-based algorithm is used to identify beam like structures by generating a curve skeleton. The curve skeleton consists of nodes, branches (consisting of one or multiple beams), and end points of branches. For each beam of the curve-skeleton, a comparative radius of the cross section at that section of the topology optimization is identified. Next, the curve skeleton is normalized by substituting each branch consisting of multiple beams by one beam. This beam has a ra-

*Corresponding author: tim.roever@tuhh.de

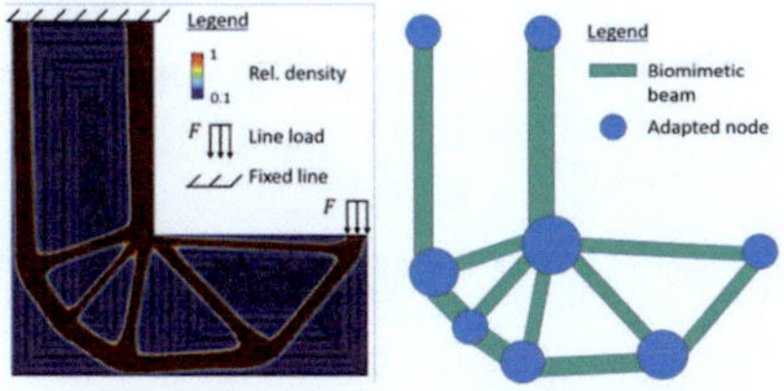

(a) Exemplary input (b) Schematic output

FIGURE 1: SCHEMATIC OF THE PROPOSED APPROACH TO CONVERT A RESULT FROM DENSITY-BASED TOPOLOGY OPTIMIZATION INTO A COMPONENT COMPOSED OF BIOMIMETIC BEAMS AND ADAPTED NODES

dius of the average of all comparative radii of the cross sections of all shorter beams of the respective branch. By that, a beam like CAD model can be generated consisting of cylindrical beams and preserved regions from the original topology optimization. Finally, a validation is done by creating a mixed dimensional finite element model. The normalized beams are simulated as 1D finite element beams while preserved regions are simulated using 3D finite elements. By applying the original load case to this model, internal stresses in all (normalized) beams can be evaluated. Based on the occurring stresses and the yield strength of the material, it is decided whether the design can support the loads or not.

In case the design does not fulfill the stress objective, two solutions are proposed. The first one is to modify the beams in which the stress objective is not achieved. The respective beams can be optimized such that they are supporting the loads sufficiently. The second option is to rerun the whole methodology with a topology optimization with an increased volume fraction. In case the newly generated design achieves the stress objective, the design is accepted as a final design. In case the newly generated design does not achieve the stress objective, another iteration is gone through with an even higher volume fraction. The number of iterations and therefore also the volume fraction are increased until the generated design is accepted based on the stress objective.

As the two methodologies [4, 5] are based on a skeletonization of a topology optimization result, a literature review was done in the fields of computer graphics, computational geometry, and visualization to derive so-called curve skeletons from two dimensional and three dimensional bodies. Cornea *et al.* [6] classified numerous methodologies for curve skeleton extraction into the four groups: thinning and boundary propagation, distance field based, as well as geometric and general field functions. The authors compared the four general approaches and found that the general field function approach gives the cleanest and smoothest results. At the same time, they also identified that this method resulted in the highest computational cost.

Because the methodologies [4, 5] use cylindrical beams and spherical nodes, the authors of this article expect that the methodology presented in this work can generate component designs of reduced mass while supporting loads sufficiently by introducing biomimetic beams and topology-optimized nodes.

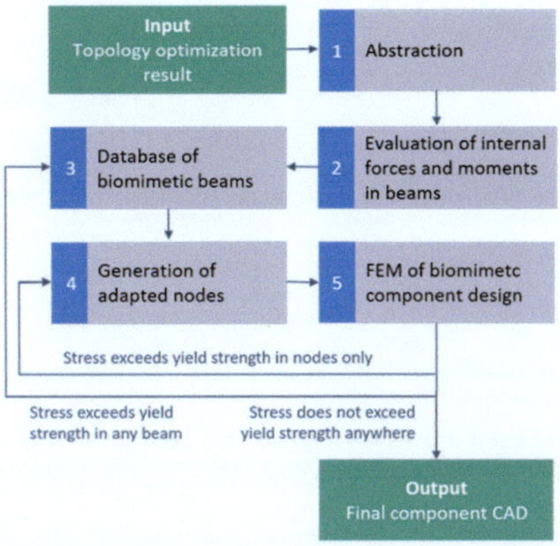

FIGURE 2: FIVE-STEP METHODOLOGY FOR THE INTEGRATION OF BIOMIMETIC BEAMS IN ABSTRACTED TOPOLOGY OPTIMIZATION RESULTS

3. PROPOSED METHODOLOGY

A schematic of the proposed approach is shown in Fig. 1. It is assumed that all regions of a mechanical topology optimization result can be classified into one of two groups, namely, 'beam-like structures' and 'nodes'. The methodology creates an auxiliary abstraction of the topology optimization result, identifies internal forces and moments in each beam like structure in the auxiliary model, to then substitute the beam like structures by biomimetic beams in order to obtain a more efficient design in terms of mass and structural integrity. One of the strengths of the proposed methodology compared to [4, 5] is that for each beam, internal forces (normal force and shear force), internal torsion, and internal bending moments are identified such that each beam can be substituted by a biomimetic beam suitable for the corresponding load case. The authors suggest that the proposed methodology leads to designs of lower mass while supporting defined load cases sufficiently.

Figure 2 gives an overview over the five main steps of the methodology. In step 1, all material allocated by the topology optimization is classified as being part of either nodes or beams, which is denoted as skeletonization. Location and orientation of these entities are derived from the topology optimization result. In step 2, the internal forces and moments in all beams are identified with respect to both of their ends. In step 3, the load case of each beam is given as an input to a database of biomimetic beams. The result is a suitable biomimetic beam as an output for every input load case. In step 4, adapted nodes are designed to

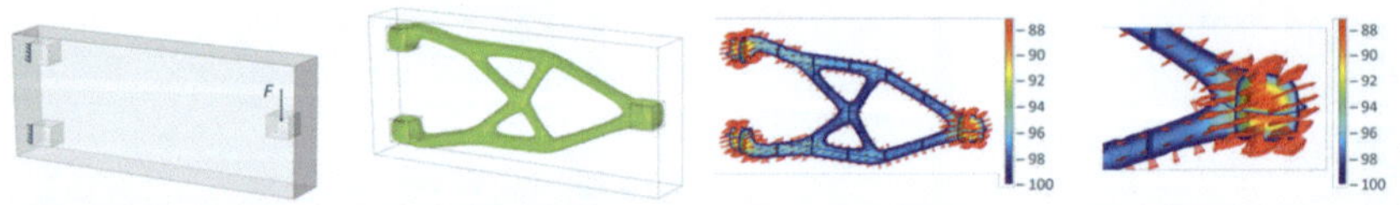

(a) Schematic of design space and boundary conditions (b) TO result showing relative densities above 0.7 (c) Electrical field within the TO result (d) Detail view of electrical field within the TO result

(e) Streamline plot of trajectories of particles in the electrical field (f) Detail view of streamline plot from (e) (g) Abstraction of the TO result by nodes and lines (h) Auxiliary model for the proposed methodology

FIGURE 3: EXEMPLARY CASE FOR THE ABSTRACTION STEP OF THE PROPOSED METHODOLOGY (IN (C) AND (D), THE ELECTRICAL POTENTIAL IN VOLTS IS GIVEN BY SCALE AND ARROWS SHOW FIELD DIRECTIONS)

connect the biomimetic beams and preserved regions to generate an intermediate biomimetic component design. And in step 5, a design iteration and a validation of the final design are performed. Finally, a CAD file is generated for the overall design of the component.

Each step of the methodology is discussed in detail in the following sections. It is noted that the methodology is not restricted to 2D problems as sketched in Fig. 1 but is also valid in 3D.

3.1 Abstraction

The goal of the input topology optimization result abstraction is to identify 'beams' and 'nodes' in the optimization result and generate an auxiliary design using spheres (nodes) and cylinders (beams). For the presented methodology, it is assumed that all regions of high density of the results can be assigned to one of the two groups. Relevant information to be identified is:

(a) coordinates of nodes,
(b) information about connections of nodes (also initial abstraction of beams),
(c) average diameter of each beam.

Inspired by the methodologies from [4, 5] as well as the curve skeletonization methods from [7, 8], a methodology for abstraction of the topology optimization result is proposed. It is an adapted version of the potential field approach and, therefore, is assumed to have better accuracy in terms of identification of nodes than the methodologies from [4, 5].

In Fig. 3(a), the design space and mechanical boundary conditions of an exemplary case are depicted. Figure 3(b) shows the respective 3D topology optimization result, which is a possible input for the proposed abstraction methodology. In a first step, a negative electric potential is added to the surface of the volume and a positive space charge density is added to the volume. This results in an electric field with gradients oriented from the center of nodes to the surface of the model, Figs. 3(c) and (d).

In a second step, particles of negative charge are spread on the surface of the model. In a time-dependent study, the particles are exposed to the electric field and an additional friction force is added such that the particles move along within the electric field and accumulate in the various nodes of the volume. The friction force helps to achieve a steady state of the particles faster by reducing oscillations around the nodes. Figures 3(e) and (f) show streamline plots of the paths that the particles moved along the electric field. The final locations of all particles are very close to each other in the center of each node.

In a third step, an algorithm is run to identify the nodes. For the final position of each particle, it is evaluated which other particles are within a certain maximum distance. Particles that are within this distance are assigned to the same node. Average values for the x, y, and z coordinates of all particles assigned to one node are defined as the coordinates of the node. Then, connections between nodes based on the original model are identified using another algorithm. For every particle in its initial position, neighboring particles are identified by a maximum distance. If two neighboring particles move into separate nodes during step 2, the corresponding nodes are considered connected by a beam. Thus, by using the obtained information, an abstraction of the initial 3D volume model can be generated using lines and nodes, Fig. 3(g). Algorithms discussed in this paragraph were developed by the authors and implemented in COMSOL Multiphysics using Java code.

To obtain data on the thickness of the beam like structures of the input topology optimization, trajectories of the particles can be evaluated. Looking at the beam like structures, particles are initially at the surface of the topology optimization result and move from there along the electric field toward the center of the beam like structure. After they reached the center of the beam like structure, they move toward the nodes of the topology optimization result along the center curve of the beam like structures. It is noted that this is an idealized model regarding the trajectories of the particles. As the trajectories show a sudden change of direction of the movement after they reached the center of the beam like structure, this characteristic can be used to distinguish each trajectory into two parts:

1. path from starting point of particle to center of beam like structure,
2. path from center of beam like structure (end of path 1) until node is reached.

Based on common type 2 paths of particles and knowledge about interconnection of nodes, particles of the same beam like structure can be identified. Using this information based on paths 1 of each particle, an approximation on the average thickness of each of the beam like structures can be obtained. It is noted that this methodology has a certain similarity with the method used for evaluation of thickness of beam like structures used in [4]. Particles' paths close to nodes are excluded from evaluation using a distance filter. Although this approach uses an idealized model of the trajectories, it is expected that the quality of the results is sufficient for the overall methodology.

Next, an auxiliary 3D model is generated as an abstraction of the original topology optimization result by modeling nodes as spheres and beams as cylindrical beams. The radii of the cylindrical beams are based on the previously identified average thicknesses of beam like structures of the topology optimization result. Finally, preserved regions (if existing) that were defined for the input topology optimization are added as copies to the model and are united by a Boolean operation to obtain the auxiliary model, Fig. 3(h). Partial implementation of the methodology as shown in Fig. 3 for presentation in this article was done in COMSOL Multiphysics 5.4.

3.2 Evaluation of internal forces and moments in beams

For evaluation of internal forces and moments in the beams, the auxiliary model is exposed to the original boundary conditions in a finite element analysis (FEA). Figure 3(h) shows the auxiliary model that can be used. Based on the FEA results, the internal forces and moments at both ends of each of the beams are evaluated.

It can be assumed that for each beam, normal forces, shear forces, and torsion (moment around the beam's main axis) are constant along the beam's length, as forces, torsion, and bending moments only act at the ends of the beams where the nodes are. Bending moments in the beams are assumed to be composed of two parts. The first part is constant and, due to a bending moment, acting at the ends of the beam. The second part is a bending moment due to shear forces at the ends of the beams that increases with increasing distance from the node.

Cut surfaces are generated at 20 % and 80 % of the length of each beam. At these cross sections, internal forces, torsion, and bending moments are evaluated. Bending moments at 20 % and 80 % of the lengths of the beams are used for extrapolation such that bending moments at 0 % and 100 % of the lengths of the beams can be obtained. The authors expect that cut surfaces should have a certain distance from nodes such that complex stress distributions at nodes do not interfere with the evaluation of internal forces and moments based on the assumptions. At the same time, the two cut surfaces in a beam should have a certain distance between them to ensure high quality results for extrapolation toward the ends of the beams. Evaluation at 20 % and 80 % was chosen as a compromise for the presented methodology. For all other obtained quantities, it is checked whether the following is true (for the nomenclature, see Fig. 4):

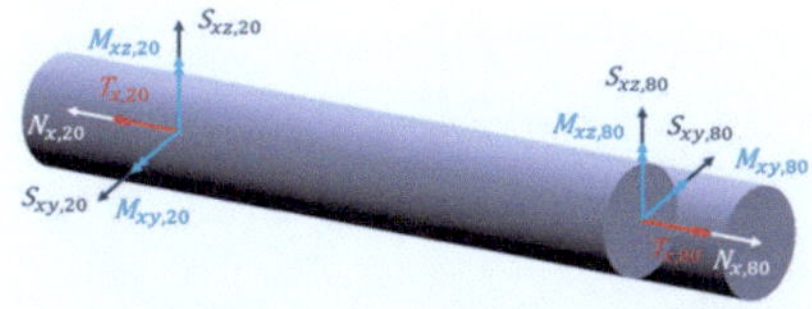

FIGURE 4: INTERNAL FORCES, TORSION, AND BENDING MOMENTS AT CUT SURFACES AT 20 % AND 80 % OF THE LENGTH OF THE BEAM WITH INTERNAL NORMAL FORCES ($N_{x,20}$, $N_{x,80}$), SHEAR FORCES ($S_{xy,20}$, $S_{xz,20}$, $S_{xy,80}$, $S_{xz,80}$), TORSIONS ($T_{x,20}$, $T_{x,80}$), AND BENDING MOMENTS ($M_{xy,20}$, $M_{xz,20}$, $M_{xy,80}$, $M_{xz,80}$)

$$N_{x,20} \approx N_{x,80} \, , \quad (1)$$

$$S_{xy,20} \approx S_{xy,80} \, , \quad (2)$$

$$S_{xz,20} \approx -S_{xz,80} \, , \quad (3)$$

$$T_{x,20} \approx T_{x,80} \, . \quad (4)$$

If the above is true, the calculated values are used. A schematic of a beam with internal forces, moments and torques is shown in Fig. 4.

Therefore, a normal force (N_x), two shear forces components (S_{xy} and S_{xz}), a torque (T_x) and two bending moment components (M_{xy} and M_{xz}) have to be evaluated at numerous cross sections of beams in the auxiliary Finite Element Method (FEM) model. The expressions for evaluation of internal forces and moments in the cut surfaces are derived from basic mechanics.

3.3 Database of biomimetic beams

Based on the loading case, each beam may be substituted by a biomimetic beam. It is expected that biomimetic beams may increase the stiffness of the structure or, in turn, reach a similar stiffness at a lower mass. Emmelmann *et al.* [9] suggests that use of biomimetic beams in structural parts can improve the mechanical performance of the overall component. Ma *et al.* [3] showed that biomimetic beams based on geometric characteristics of bamboo can have a considerably increased load bearing capacity for certain load cases.

An exemplary biomimetic beam is shown in Fig. 5. In this case, the cut surface of a trunk of a banana plant was abstracted, Fig. 5(b). Extrusion of the abstraction leads to a biomimetic beam, Fig. 5(c). Depending on its length, the beam can support certain load cases.

For the presented design methodology, a database of biomimetic beams is proposed. Figure 6 shows the function of the database. Its inputs are the length of the beam and its normal forces, shear forces, torsions, and bending moments at both ends of the beam.

Various parametric biomimetic beams are stored in the database. For each of the parametric beams, numerous versions with different parameters are included (lines in simplified table in Fig. 6). Furthermore, a function is provided with respect to

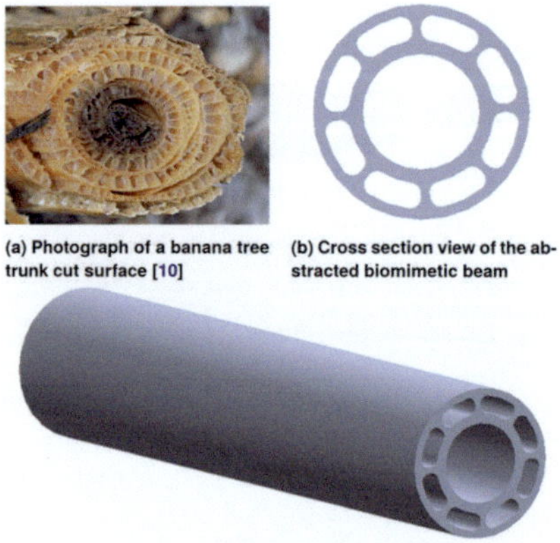

(a) Photograph of a banana tree trunk cut surface [10]

(b) Cross section view of the abstracted biomimetic beam

(c) Isometric view of the biomimetic beam

FIGURE 5: EXEMPLARY BIOMIMETIC BEAM BASED ON A BANANA TREE TRUNK

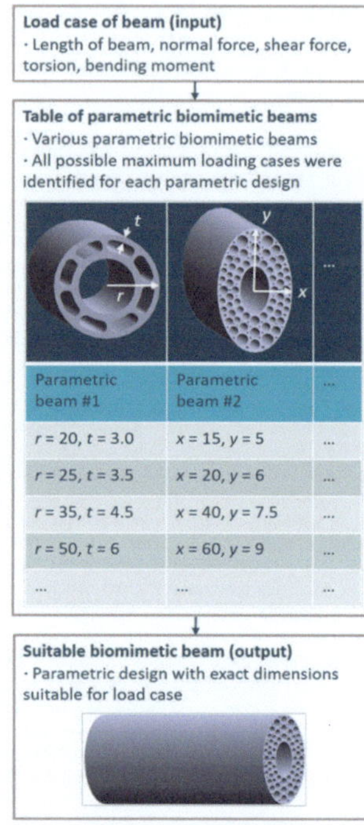

FIGURE 6: SCHEMATIC OF THE BIOMIMETIC BEAM DATABASE UTILIZATION

normal forces, shear forces, torsions, and bending moments evaluating the maximum load cases that this specific type of beam can successfully support.

With a given load condition of a beam from the auxiliary model, the following steps have to be carried out. First, the best suitable specific design of the parametric beam #1 is identified. Starting with the specific design of the smallest cross sectional area and carrying on with the specific designs that have greater cross sectional areas, it is checked whether the design can support the given load. The first specific design that is found to successfully support the loads is identified as the best suitable specific design of parametric beam #1. Analogously, for parametric beams #2, #3, etc., the most suitable specific types of designs are identified. Finally, the mass of all identified specific designs is compared and the design with the smallest mass is chosen. This specific design is the output of the digital database for the given input load condition.

3.4 Generation of adapted nodes

Based on the abstraction of the original topology optimization results by lines and nodes [Fig. 3(g)] and the output from the database of biomimetic beams, the orientations and locations of biomimetic beams in 3D space are known. With this information, adapted nodes can be generated to connect the biomimetic beams with each other as well as the biomimetic beams and the preserved regions of the input topology optimization. In the following, two approaches are presented for the generation of these adapted nodes.

In the first approach, all biomimetic beams and connecting surfaces are positioned in a topology optimization model. The bounding box of all components in the model is used as a design space for the topology optimization. All connecting surfaces and biomimetic beams are fixed at a relative density of one. All mechanical boundary conditions of the original input topology optimization are added to the new model. The optimization objective is defined using the minimum mass approach in which only as much material as needed is allocated to support the loads without the material reaching a certain threshold, such as 80 % of the yield strength of the material at any point inside the component.

As the biomimetic beams can be expected to support loads that act at their ends, material should be placed by the algorithm at nodes only. Figure 7 shows a conceptual representation of nodes generated by the presented approach in a 2D example. To generate the shape of the nodes and also the whole biomimetic model, a filter can be used to only export all volume where the relative density exceeds a certain threshold (e. g., 0.7).

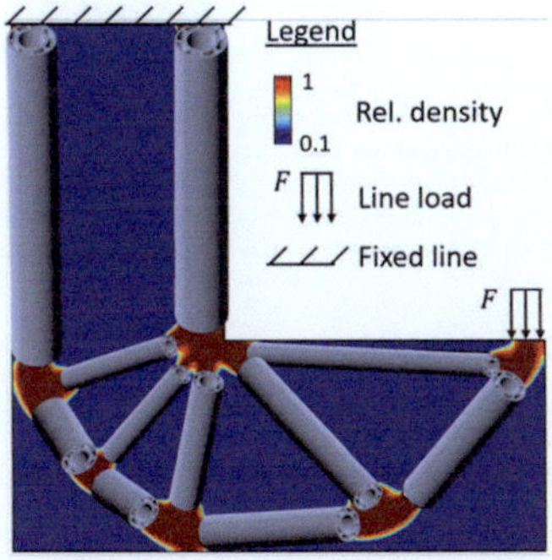

FIGURE 7: CONCEPTUAL REPRESENTATION OF GENERATION OF NODES FOLLOWING THE FIRST APPROACH FOR AN EXEMPLARY 2D COMPONENT (RELATIVE DENSITIES OF THE RESPECTIVE DESIGN SPACE ARE GIVEN BY SCALE)

In the second approach, individual topology optimization models are generated for each node. Based on the abstraction of the original topology optimization results by lines and nodes [Fig. 3(g)] and the output from the database of biomimetic beams, a topology optimization model such as the one in Figure 8 can be generated. The design space contains the ends of all biomimetic beams that are to be connected by the node.

The topology optimization model is set up the following way: Biomimetic beams that are to be connected by the node are placed into the model. However, only the ends of the biomimetic beams are added rather that the whole biomimetic beams. Internal forces and moments identified with the auxiliary model are used to introduce forces and moments as boundary conditions to the ends of the biomimetic beams in the new topology optimization. In case preserved regions are within a certain distance to the node, they are added to the model as well as loads as defined in the input topology optimization of the overall methodology.

Analogously to the first approach, all preserved regions and biomimetic beams are fixed at a relative density of one. Furthermore, also as in the first approach, the minimum mass approach is used (see above). To generate the shape of the node, all volume in which the relative density exceeds a certain threshold (see above) can be exported from each of the models. Using this methodology, shapes for all nodes can be generated.

In a final step, biomimetic beams, generated nodes, and preserved regions can be positioned and oriented in 3D space and united in a single model.

In the following, the two approaches are briefly discussed. The first approach is a holistic approach. It can be expected that the result of this approach supports all loads of the original mechanical problem. However, to follow this approach, relatively large stiffness matrices need to be set up and solved, as the geometry of the biomimetic beams is complex. Furthermore, when

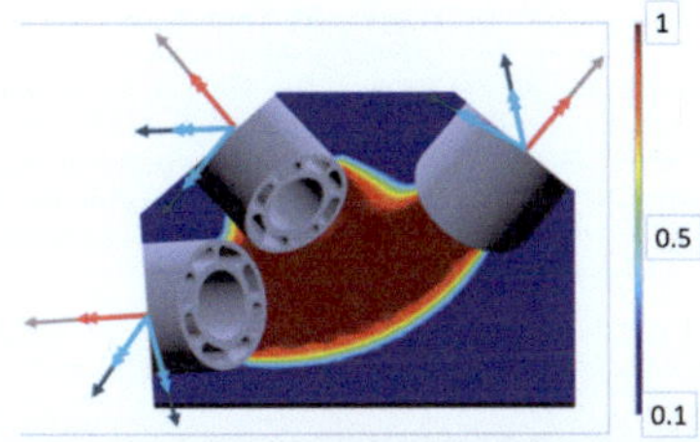

FIGURE 8: CONCEPTUAL REPRESENTATION OF GENERATION OF NODES FOLLOWING THE SECOND APPROACH FOR AN EXEMPLARY 2D COMPONENT (RELATIVE DENSITIES OF THE RESPECTIVE DESIGN SPACE ARE GIVEN BY SCALE)

applying topology optimization algorithms, usually fine meshes are needed to generate designs of high mechanical performance. Therefore, rather high computational costs are expected for this approach.

The second approach subdivides the problem of node generation in multiple problems. This reduces the size of the stiffness matrices to be used tremendously or, in turn, increases the quality of the result as the volume at the node can be discretized by a much finer mesh. The downside of this approach is that the forces and moments acting inside the biomimetic beams can be expected to have a certain error. This is due to the fact that substitution of the beams in the auxiliary model by biomimetic beams leads to changed force paths and moments in the structure. Therefore, the nodes generated using this approach may be oversized or undersized. Future research has to be done to investigate advantages and disadvantages of the two approaches.

3.5 FEM of biomimetic component design and output

Based on the previously described methodologies, volumetric models of new biomimetic component designs containing biomimetic beams can be generated. In this step, the original boundary conditions are applied to the new design in an FEA. In post processing of the FEA results, all areas in which the von Mises stress exceeds the yield strength of the material are identified. According to Fig. 2, three different cases are to be distinguished:

- If there are no areas in which the yield strength is exceeded by the von Mises stress, the component design is accepted as the final design.
- If the yield strength is exceeded by the von Mises stress in any of the nodes, the step of node generation is carried out again, using either smaller finite elements or adapting the threshold of the optimization objective (only as much material as needed is allocated to support the loads without the material reaching, e. g., 70 % instead of 80 % of the yield strength of the material at any point inside the component). With the achieved design, the workflow is carried on from step 5 'FEM of biomimetic component design'.

- If the yield strength is exceeded by the von Mises stress in any of the beams, the step utilizing the database of biomimetic beams is partly carried out again. For all beams in which the yield strength is exceeded, previously chosen biomimetic beams are substituted by the type of this parametric beam that is next in terms of structural integrity. With the achieved results, the workflow is carried on from step 4 'Generation of adapted nodes.'

Using this methodology, a final biomimetic component design can be generated that meets the defined stress criterion.

4. CONCLUSION AND OUTLOOK

Although structural component designs combining topology optimization and incorporation of biomimetic beams seem to have great potential, they are not widely used. For this reason, a five step design methodology was proposed in this article in which topology optimization results are converted into new component designs including biomimetic beams leading to designs of lower mass while sufficiently supporting the acting loads. The methodology has high potential to allow for reproducible biomimetic component designs, a trackable and easily documentable component development process, and the possibility of automating the design process to ultimately save development costs when designing structural components.

As the methodology was developed on a conceptual level, detailed implementation, testing, and comparison to multiscale topology optimization approaches is still to be done and will be addressed by the authors in future works.

REFERENCES

[1] Zhu, J.-H., Zhang, W.-H. and Xia, L. "Topology optimization in aircraft and aerospace structures design." *Archives of Computational Methods in Engineering* Vol. 23 No. 4 (2016): pp. 595–622.

[2] Yang, R. J. and Chahande, A. I. "Automotive applications of topology optimization." *Structural Optimization* Vol. 9 No. 3 (1995): pp. 245–249.

[3] Ma, J., Chen, W., Zhao, L. and Zhao, D. "Elastic buckling of bionic cylindrical shells based on bamboo." *Journal of Bionic Engineering* Vol. 5 No. 3 (2008): pp. 231–238.

[4] Nana, A., Cuillière, J.-C. and François, V. "Automatic reconstruction of beam structures from 3D topology optimization results." *Computers & Structures* Vol. 189 (2017): pp. 62–82.

[5] Cuillière, J.-C., François, V. and Nana, A. "Automatic construction of structural CAD models from 3D topology optimization." *Computer-Aided Design and Applications* Vol. 15 No. 1 (2018): pp. 107–121.

[6] Cornea, N. D., Silver, D. and Min, P. "Curve-skeleton properties, applications, and algorithms." *IEEE Transactions on Visualization and Computer Graphics* Vol. 13 No. 3 (2007): p. 530.

[7] Ahuja, N. and Chuang, J.-H. "Shape representation using a generalized potential field model." *IEEE Transactions on Pattern Analysis and Machine Intelligence* Vol. 19 No. 2 (1997): pp. 169–176.

[8] Cornea, N. D., Silver, D., Yuan, X. and Balasubramanian, R. "Computing hierarchical curve-skeletons of 3D objects." *The Visual Computer* Vol. 21 No. 11 (2005): pp. 945–955.

[9] Emmelmann, C., Petersen, M., Kranz, J. and Wycisk, E. "Bionic lightweight design by laser additive manufacturing (LAM) for aircraft industry." *SPIE Eco-Photonics 2011: Sustainable Design, Manufacturing, and Engineering Workforce Education for a Green Future*, Vol. 8065: pp. 181–192. 2011.

[10] foxtail_1. "banana 1." https://wordpress.org/openverse/photos/4a31b903-ab5b-4485-9a97-3f1b4d65028c. Licensed under CC BY-NC-SA 2.0.

A.2 Development and assessment of a methodology for abstraction of topology optimization results to enable the substitution of optimized beams

The following pages feature a reprint of the journal article "Development and assessment of a methodology for abstraction of topology optimization results to enable the substitution of optimized beams," originally published in 2023. This reprint is provided under the terms of the CC BY 4.0 license.

Journal of Laser Applications

RESEARCH ARTICLE | NOVEMBER 03 2023

Development and assessment of a methodology for abstraction of topology optimization results to enable the substitution of optimized beams

Special Collection: Proceedings of the International Congress of Applications of Lasers & Electro-Optics (ICALEO 2023)

Tim Röver ; Maximilian Bader; Karim Asami ; Claus Emmelmann ; Ingomar Kelbassa

J. Laser Appl. 35, 042061 (2023)
https://doi.org/10.2351/7.0001185

Journal of Laser Applications ARTICLE pubs.aip.org/lia/jla

Development and assessment of a methodology for abstraction of topology optimization results to enable the substitution of optimized beams

Cite as: J. Laser Appl. 35, 042061 (2023); doi: 10.2351/7.0001185
Submitted: 17 July 2023 · Accepted: 5 October 2023 ·
Published Online: 3 November 2023

View Online Export Citation CrossMark

Tim Röver,[1,a)] Maximilian Bader,[1] Karim Asami,[1] Claus Emmelmann,[1] and Ingomar Kelbassa[2,3]

AFFILIATIONS

[1]Institute of Laser and System Technologies (iLAS), Hamburg University of Technology (TUHH), Harburger Schloßstraße 28, 21079 Hamburg, Germany
[2]Fraunhofer Research Institution for Additive Manufacturing Technologies IAPT, Am Schleusengraben 14, 21029 Hamburg, Germany
[3]Institute for Industrialization of Smart Materials (ISM), Hamburg University of Technology (TUHH), Eißendorfer Straße 40, 21073 Hamburg, Germany

Note: Paper published as part of the special topic on Proceedings of the International Congress of Applications of Lasers & Electro-Optics 2023.
[a)]**Electronic mail:** tim.roever@tuhh.de

ABSTRACT

Improving mechanical topology optimization (TO) results by substituting biomimetic beams is one possibility to achieve designs of mechanical components that are highly sustainable and show good mechanical performance. Because of their geometric complexity, such designs were found to be well-suited for production by laser additive manufacturing. One obstacle of incorporating biomimetics beams in TO designs is the lack of detailed design methodologies. Röver *et al.* ["Methodology for integrating biomimetic beams in abstracted topology optimization results," in *Proceedings of the ASME 2022 International Mechanical Engineering Congress and Exposition. Volume 4: Biomedical and Biotechnology; Design, Systems, and Complexity Columbus*, OH, 30 October–3 November (ASME, New York, 2022)] proposed a corresponding design concept. Building on their concept, we present in this work a detailed methodology for abstraction of TO results to a design consisting of ball nodes and cylindrical beams. Using such an auxiliary design, the internal forces and moments of the beams can be evaluated to allow for the substitution of suitable biomimetic beams to generate biomimetic component designs in a next step. We present a skeletonization algorithm based on the potential field approach. Using the skeletonization and an additional analysis of the dimensions of the beams in the TO result, the algorithm develops an auxiliary design of the original TO result. The final algorithm was applied to three common TO results to obtain one auxiliary component design each. The developed algorithm was found to generate abstractions that were well-suited for use in the methodology proposed in Röver *et al.* ["Methodology for integrating biomimetic beams in abstracted topology optimization results," in *Proceedings of the ASME 2022 International Mechanical Engineering Congress and Exposition. Volume 4: Biomedical and Biotechnology; Design, Systems, and Complexity Columbus*, OH, 30 October–3 November (ASME, New York, 2022)], because internal forces and moments in the abstracted beams could be evaluated with less effort. Therefore, our work contributes to a detailed design methodology for biomimetic mechanical components in the field of design for additive manufacturing.

Key words: design for additive manufacturing (DfAM), skeletonization, generative design, design optimization, biomimetics, topology optimization, finite element analysis, PBF-LB/M

Journal of Laser Applications ARTICLE pubs.aip.org/lia/jla

INTRODUCTION

Lightweight structures are becoming increasingly relevant in the automotive and aerospace industries. A lower weight ensures that less energy is needed to move, i.e., less fuel is consumed. At the same time, the use of a lightweight structural element reduces the amount of manufacturing material that is consumed. It was suggested in previous work that by using bionic structures, the weight of structural designs developed by topology optimization (TO) can be further reduced while maintaining structural stability.[1]

Biomimetic structures are inspired by nature in order to transfer the modes of operation from biology to technical applications.[2] References 1 and 3 were one of the first works in which laser additive manufacturing and biomimetics were combined to generate novel designs for structural components. In both works, aviation applications were focused.

References 4 and 5 show the potential for performance and functionality increase of components and systems when combining additive manufacturing (AM) and biomimetics for a variety of applications. For a more comprehensive and general information on the field of biomimetics, the reader is kindly referred to Refs. 6–8.

Reference 9 proposed a concept for a detailed design methodology to generate biomimetic designs for structural components based on TO results. TO is a valuable tool in the design of components. However, considering material distribution-based TO approaches, the quality of the results depends on many factors, one of which is the resolution of the finite-element mesh.[10] Because of limited computational resources, the finite-element mesh cannot be refined infinitely, due to which the geometric complexity of the TO result is limited. The design methodology proposed in Ref. 9 aims to improve existing TO results by integrating biomimetic beams and redesigning nodes to achieve a higher geometric complexity of the design. The authors of the study assume that the increased geometric complexity allows for a lower mass of the component, while ensuring its structural integrity.

The methodology proposed in Ref. 9 generates designs that are producible by AM. It consists of the following steps:

- Input: The input is a result of a mechanical TO.
- Abstraction: The input abstraction algorithm develops an abstraction of the TO results in the form of a representation by points and lines using a potential field approach. In a next step, a 3D auxiliary abstraction of the input is developed. This 3D abstraction consists of cylindrical beams and ball nodes.
- Evaluation of internal forces and moments in beams: The internal forces and moments at cross sections at 20% and 80% of the length of the cylindrical beams of the auxiliary abstraction are evaluated.
- Database of biomimetic beams: For each of the cylindrical beams of the auxiliary abstraction, a biomimetic beam is developed. Development of the biomimetic beams is based on a database of beam designs and the information regarding internal forces and moments of the beams at the two analyzed cross sections and the beam's lengths.
- Generation of nodes: One of two approaches is used to develop a node for the new biomimetic component.
- Finite-element analyses of biomimetic component design: Biomimetic beams and the developed nodes are combined to achieve a complete component design. Finite-element analyses are used with possible returns to earlier steps of the methodology to ensure that the new design can support the expected loads.
- Output: The output is a design of a biomimetic component.

This work presents a detailed solution to the concept regarding abstraction proposed in Ref. 9. Moreover, it exemplarily shows how the outputs from the solution can be used for the evaluation of internal forces and moments in the beams of the abstraction.

MATERIALS

Implementation of the work was primarily done in COMSOL Multiphysics version 5.4.[11] Only the evaluation of internal forces and moments was done in COMSOL Multiphysics version 6.0.[12]

Structural steel was considered as the material of all mechanical structures in this work. Material parameters were obtained from the Comsol Multiphysics Material Library and are given in Table I.

ABSTRACTION METHODOLOGY

In the following section, the developed abstraction methodology is presented. First, the development of three different input TO results for the application of the abstraction methodology is presented. Next, the abstraction methodology itself is presented in detail. Figure 1 gives an overview over the respective steps. In the last subsection, the evaluation of internal forces and moments in the beam cross sections of one of the 3D abstractions is demonstrated. For more detailed information on the implementation of the abstraction methodology, the reader is kindly referred to the supplementary material,[13] which contains abstraction models of the Michell cantilever.

Input topology optimizations

In Figs. 2(a)–2(c), the design spaces, their dimensions, mechanical constraints, and the final TO results of three mechanical problem formulations are depicted. These TO results were developed as an input for the abstraction methodology of this work. In the following paragraphs, the generation of these results is briefly presented.

All three TO simulations are well known and often used as exemplary cases in the field of TO. The detailed problem design used in this work was based on Ref. 14. The three problems are referred to as Michell cantilever, L-shape, and Messerschmitt-Bölkow–Blohm (MBB) beam.

The Michell cantilever's design space has a length of 180 mm, a height of 60 mm, and a thickness of 10 mm. The cantilever is supported by a fixed clamping on one side. On the other side, a small surface is subjected to the mechanical load of 500 N in the

TABLE I. Material properties as considered in numerical analyses.

Property/alloy	Structural steel
Elastic modulus (GPa)	200
Poisson's ratio (–)	0.30
Density (kg/m^3)	7850

Journal of Laser Applications ARTICLE pubs.aip.org/lia/jla

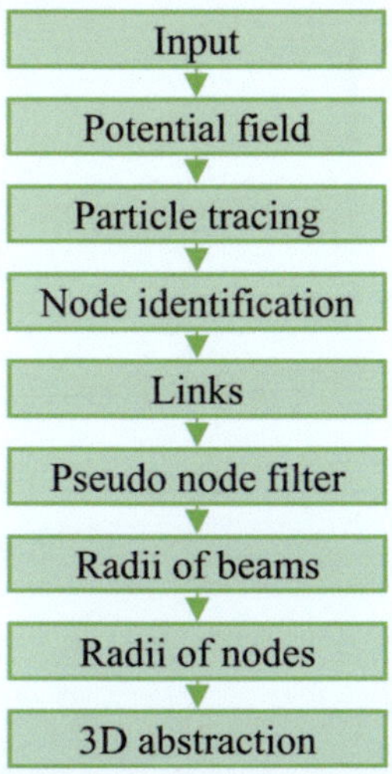

FIG. 1. Overview of the steps of abstraction methodology.

negative z-direction. The respective surface has a height of 6 mm and a width of 10 mm.

The L-shape's design space length of the long sides is 200 mm. The short sides have an edge length of 120 mm. A radius was added to the notch of the component in order to avoid stress peaks. The radius has a value of 20 mm. The model has a thickness of 10 mm. The component is mechanically fixed at the top surface. An area for load application was defined with a length of 20 mm. The load applied in the negative z-direction was defined as 600 N.

The MBB beam's design space has a length of 200 mm, a height of 40 mm, and a thickness of 10 mm. Mechanical boundary conditions for this model were partly based on Ref. 15. The bearings consist of a fixed clamping and a roller bearing. These are defined on the lower level of the model at each end. On the top side, an area of 10 mm length was created for load application. As before, the applied force is oriented in the negative z-direction and in this case has a value of 1000 N.

Hexahedral finite-element (FE) meshes were generated for all three TOs using the swept meshing function implemented in Comsol Multiphysics 5.4.[11] Maximum and minimum element sizes were chosen equally for the models and had values of 1 mm (Michell cantilever), 1.5 mm (L-shape), and 1 mm (MBB beam) based on Ref. 14.

The density-based solid isotropic material with penalization (SIMP) method, as implemented in Comsol Multiphysics 5.4[16] based on Refs. 17 and 18, was used to generate the TO results. In addition to the mechanical boundary conditions, an integral objective was implemented at all six outer surfaces of the TO models to hinder material to be distributed close to the outer areas of the design spaces. This simplifies the export of the TO results in the used software.

Relevant parameters of the TOs of this work are given by the penalty factor $p = 6$ and the volume fraction $v_f = 0.25$. The models were optimized in 70 optimization iterations.

Abstraction steps

The abstraction methodology was developed based on the concept proposed in Ref. 9, which itself was based on Refs. 19–22. Its output is a 3D component design that is an abstraction of the input TO result. The abstraction consists of cylindrical beams and ball nodes. The size of the cylindrical beams and ball nodes is based on the input TO result. It is noted that the abstraction is strongly connected to the research on curve skeletons. For a comprehensive overview of the topic, the reader is kindly referred to Ref. 23.

In the following subsections, the steps of the abstraction methodology are presented. An overview of the abstraction methodology is given by input, output, and selected interim results, as depicted in Fig. 3.

Input

Input for the developed abstraction methodology is given by TO results such as the ones depicted in Figs. 2(a)–2(c) and 3(a). They consist of 3D solids.

Potential field

A negative electric potential (V_0) is added to the surface of the TO design and a positive space charge density (ρ_v) is applied to the volume of the same. A static numerical simulation results in an electric field within the 3D volume with a minimal potential at the model surface and increasingly higher potentials toward the local cores. Therefore, the gradients are oriented from the "inside" to the "outside." An exemplary result of this step can be found in Fig. 3(b).

Particle tracing study

Building on the previous step, a certain number of particles (n_P) of negative charge are spread equally on the surface of the 3D volume. Figure 3(c) shows the particles (green) in their initial positions. These particles are subject to the electric field. Furthermore, a friction force with a background number density (N_d) is added to the study of moving particles. In a time-dependent numerical analysis, the particles then accumulate in various points within the 3D volume [cf. Fig. 3(e)]. The accumulation locations can be attributed to two groups. One group of particles accumulates close to the nodes of the TO design. The other group of particles grinds to a halt inside the beam-like structures that connect nodes. It is noted that the friction force ensures that the particles grind to a halt and prevent their oscillatory motion. The prevention of oscillation of

Journal of Laser Applications ARTICLE pubs.aip.org/lia/jla

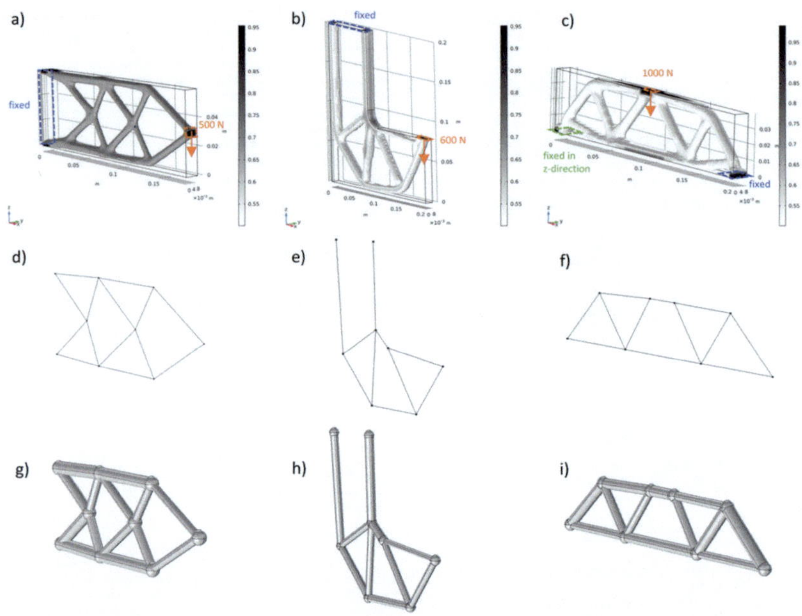

FIG. 2. (a)–(c) TO results developed as input for abstraction methodology. Relative densities are given by the scales. (d)–(f) Abstraction of TO results by lines and points, which are the interim results of the presented methodology. (g)–(i) 3D Abstractions of TO results by cylindrical beams and ball nodes.

particles is of high importance to ensure high-quality abstraction results.

Node identification

In a next step, accumulations of particles are identified. This is done by investigating the distances between the locations of the particles at the end of the time-dependent study. Starting with the first particle, all other particles that are within a defined small radius ($r_{\text{nodes\_filter}}$) are attributed to the same node. Moving on with the second particle, it is checked whether or not the particle was already attributed by the algorithm. If it was already attributed, the algorithm moves to the next particle. If it was not attributed, its final coordinates are defined as a new node and again all other particles that are within the $r_{\text{nodes\_filter}}$ are again attributed to this node. Therefore, a long list of the particles' final coordinates at the end of the time-dependent study is reduced to a list of nodes' coordinates.

Links

In a next step, connections between nodes are identified by analyzing the particles' initial locations at the beginning of the time-dependent study and their attribution to the nodes based on the previous step. A threshold $r_{\text{links\_filter}}$ is introduced. All possible initial particle locations of two particles are analyzed. If the distance between two initial locations of the two particles is within the threshold $r_{\text{links\_filter}}$ and both particles are attributed to two different nodes, a link is successfully identified. Figure 3(f) depicts an exemplary result after this step.

Journal of Laser Applications ARTICLE pubs.aip.org/lia/jla

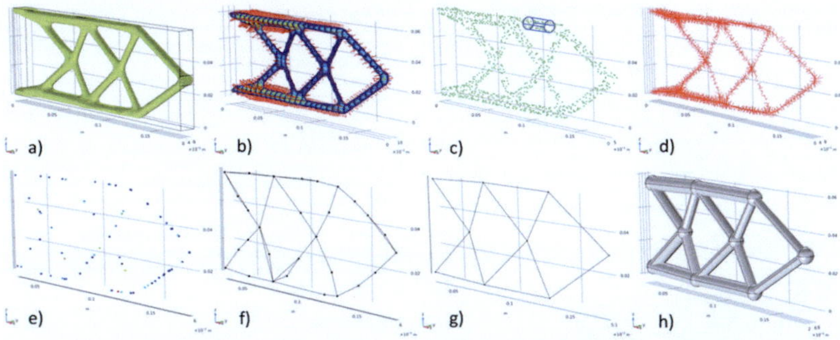

FIG. 3. Overview of the input, output, and interim results of the abstraction methodology.

Pseudo node filter

Next, a filter algorithm is applied to improve the interim abstraction result. One after another, all nodes of the interim results are analyzed that have one or two nodes connected to them by links. The angle between the two linking lines that connect the three nodes is analyzed. If the angle α is $153° < \alpha < 190°$, the center node is classified as a pseudo node. The pseudo node and its links are deleted and a direct link between the remaining two nodes is introduced. After all nodes are analyzed as previously explained, the same procedure is run once more for all remaining nodes to remove pseudo nodes that might have developed after the first filter cycle.

After filtering pseudo nodes using the presented method, an abstraction consisting of lines and points based on the true nodes and their links can be generated [cf. Fig. 3(g)].

Radii of beams

For generating 3D abstraction values, the radii of the cylindrical beams and ball nodes need to be defined. For each link connecting two nodes, the following is done: the linking line segment connecting the two nodes is reduced by 20% of its length at both ends. Then, a hypothetical cylinder is generated to be used as a probe. Figure 3(c) shows an exemplary cylindrical probe used for identifying the thickness of one of the beam-like structures in the TO result. The cylinder's axis is given by the previously reduced line segment. The radius of the cylinder r_{cyl_probe} is defined to be 10 mm for all three models in this work. A schematic of the cylindrical probe is depicted in Fig. 3(c). Considering the starting points of the particles at the beginning of the time-dependent study, all particles that are within the cylindrical probe are identified. These points are used to obtain a value for the thickness of the beam connecting the two nodes.

First, the regression line of all starting points within the probe is identified. For implementation, the Java package[24] was used. The position vector of the regression line is identified by using the average values for the x-, y,- and z-coordinates of all starting points, respectively.

The initial locations of the particles inside the cylindrical probe are then used to identify the direction vector of the regression line. For this, the x-, y,- and z-coordinates are saved into a matrix. From each entry, the corresponding value of the position vector is then subtracted. From this matrix, the covariance matrix is calculated by multiplying it with its respective inverse. The eigenvalues and eigenvectors of this covariance matrix are then determined by applying the power method based on Refs. 25 and 26. The eigenvector corresponding to the greatest eigenvalue is then set as the direction vector of the regression line.

For calculating the beam radius, the initial locations of the particles inside the cylindrical probe, the position vector, and the direction vector of the regression line are used. The distance between each initial particle location within the probe and the regression line is calculated. The distances of all particles are then averaged and this value is set as the radius of the beam of the abstraction.

Radii of nodes

The radii of all ball nodes in the abstraction are set to be 110% of the largest radius of any cylindrical abstraction beam as identified from the previous step.

3D Abstraction

Finally, an abstraction consisting of cylindrical beams and ball nodes can be generated. The abstraction is based on the input TO result. Figures 2(g)–2(i) and 3(h) show the respective abstractions for the three exemplary input TO results considered in this study.

Journal of Laser Applications ARTICLE pubs.aip.org/lia/jla

Evaluation of internal forces and moments in abstractions

In Fig. 4, the results of evaluation of internal forces and moments in three exemplary beams of the abstracted TO result of the Michell cantilever are depicted. To achieve this result, the 3D abstraction consisting of cylindrical beams and ball nodes is subjected to the original mechanical boundary conditions of the Michell cantilever in a mechanical numerical analysis. Next, using the function of section forces as implemented in COMSOL Multiphysics 6.0,[27] the internal forces and moments at the three cross sections of the solid could be evaluated. For each cross section of a beam, the axial force (N), the shear force along the first local axis (T1), the shear force along the second local axis (T2), the twisting moment (Mt), the bending moment around the first local axis (M1), and the bending moment around the second local axis (M2) could be evaluated with low effort.

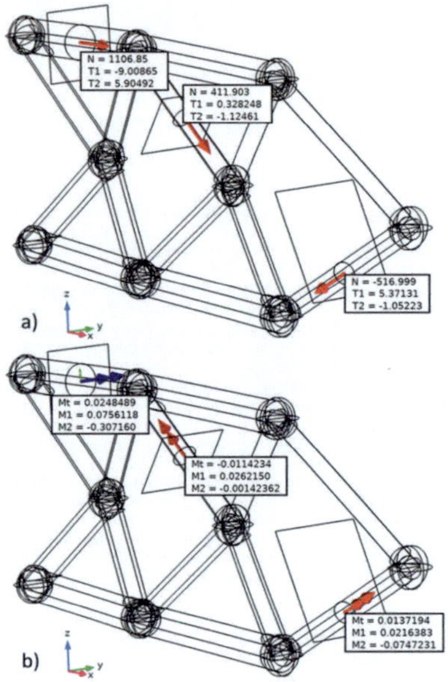

FIG. 4. Evaluation of internal forces and moments at three exemplary cut surfaces within beams of the abstracted TO result of the Michell cantilever. (a) Axial force (N), shear force along the first local axis (T1), and shear force along the second local axis (T2). (b) Twisting moment (Mt), bending moment around the first local axis (M1), and bending moment around the second local axis (M2).

Therefore, with regard to the methodology proposed in Ref. 9, it could be demonstrated that the evaluation of internal forces and moments in the beams of the 3D abstraction can be done with low effort.

The evaluation of internal forces and moments at 20% and 80% of the length of the beams, together with the information on the length of the beam that is known from the 3D abstraction, is expected to allow for the development and design of lightweight biomimetic beams according to Ref. 9 in a next step.

Hence, combining the abstraction methodology as proposed in this work with the functionality of sections forces, relevant parts of the methodology proposed in Ref. 9 were successfully developed and implemented.

DISCUSSION

In this section, the limitations and advantages of the presented work are presented and discussed.

Limitations

One limitation of the presented work is given by the choice of parameter values. Table II gives an overview over the most critical parameters of the abstraction methodology. Values for the parameters as used in this study were set based on preliminary studies on the three models. For application of the presented abstraction algorithm to other TO results, the authors expect that parameters need to be adjusted to achieve good abstraction results. A preliminary study on the impact of parameter choice on the abstraction results yielded that parameters r_{links_filter} and r_{nodes_filter} had a large impact on the quality of the abstraction result. This fact is one limitation of the work to be considered for future work on the topic.

Unbeneficial choice of parameters results in an abstraction as shown in Fig. 5. The algorithm did not filter all pseudo nodes.

TABLE II. List of the most relevant parameters of abstraction methodology.

Parameter	Explanation	Values used for Michell cantilever
r_{links_filter}	Filter radius for identification of links between nodes	7 (mm)
r_{nodes_filter}	Filter radius for identification of nodes	10 (mm)
r_{cyl_probe}	Radius of cylindrical probe for evaluation of thickness of beams	10 (mm)
n_P	Number of particles	1000
V_0	Electric potential	−100 (V)
ρ_v	Space charge density	1×10^{-6} (C/m^3)
N_d	Background number density (of friction force)	5×10^{20} (1/m^3)
α	Upper and lower threshold angles of pseudo node filter	$153° < \alpha < 190°$

Journal of Laser Applications ARTICLE pubs.aip.org/lia/jla

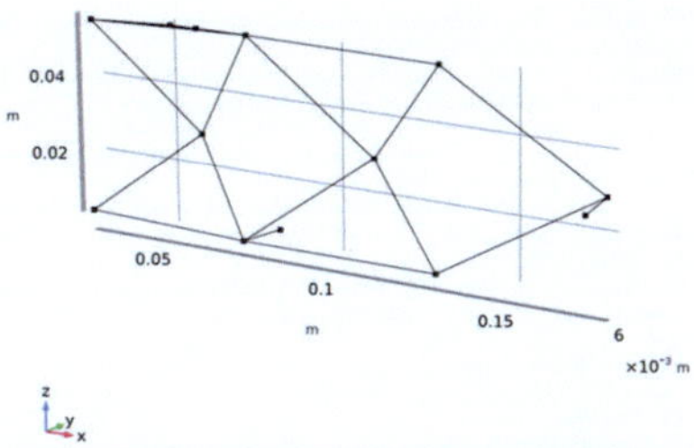

FIG. 5. Imperfect line and point abstraction of the Michell cantilever based on the use of alternative values for parameters ($r_{links\_filter} = r_{nodes\_filter} = 5$ mm).

Pseudo nodes that are linked to three other nodes are not filtered. Furthermore, pseudo nodes that have only one other node linked to them are not filtered either.

An additional limitation of the work is that the current abstraction methodology is not designed to develop abstractions for wall-like structures that might be part of the TO result. With regard to the methodology presented in Ref. 9, this is not problematic, as this methodology does not seem to focus on TO results with wall-like structures but rather seems to focus on truss-like TO results. However, if the abstraction methodology presented in this paper is to be used for another application or approach, the limitations with regard to wall-like structures should be considered.

Advantages

Implementation of the abstraction methodology in COMSOL Multiphysics has the advantage that the workflow from the setup of the TO to the evaluation of internal forces and moments in beams of the 3D abstraction can be done within one software. Further implementation of the overall methodology by Ref. 9 in COMSOL Multiphysics is expected to allow for the generation of a single executable file (EXE file) using the compiler function[28] to generate an user-friendly application that develops lightweight biomimetic component designs that are in line with design for additive manufacturing (DfAM).

This is expected to allow a widely automated design of structural components that are based on TO results as inputs. Such designs are expected to incorporate biomimetic beams to achieve a low mass, while ensuring structural integrity, and at the same time, consider the constraints by additive manufacturing techniques such as powder bed fusion of metals by a laser beam (PBF-LB/M).

SUMMARY AND OUTLOOK

A methodology for the abstraction of 3D TO results was presented. The methodology is based on previous works in the field of curve skeletons and was implemented in COMSOL Multiphysics.

Using an input TO result, a time-dependent particle tracing study is used to identify point nodes and line links for the abstraction based on the potential field approach. Filtering of the result is done for improvement of the interim result. Next, a submethodology for defining suitable dimensions of the cylindrical beams and ball nodes is applied to allow for generation of the final 3D abstraction of the input TO result.

The developed methodology was applied to three common mechanical problem formulations, and it was able to provide suitable 3D abstractions as outputs.

After applying the original mechanical loads to the 3D abstractions, the internal forces and moments at the cross sections in the beams can be evaluated with low effort. Therefore, the presented work gives detailed solutions to the first two steps of the more comprehensive design methodology for the biomimetic mechanical components proposed in Ref. 9 and contributes to the field of DfAM.

The authors plan to improve the presented work in the future by developing solutions for providing an automated choice of the relevant abstraction parameters. Furthermore, future work should focus on substitution of solid beams by biomimetic lightweight design beams and assess the generated biomimetic design with regard to lightweight characteristics and manufacturability in PBF-LB/M.

ACKNOWLEDGMENTS

Parts of this work were funded by the Fraunhofer Future AM project.

AUTHOR DECLARATIONS

Conflict of Interest

The authors have no conflicts to disclose.

Author Contributions

Tim Röver: Conceptualization (lead); Investigation (equal); Methodology (lead); Supervision (lead); Validation (equal); Visualization (lead); Writing – original draft (lead); Writing – review & editing (equal). **Maximilian Bader:** Conceptualization (supporting); Investigation (equal); Methodology (supporting); Validation (equal); Visualization (equal); Writing – original draft (equal); Writing – review & editing (supporting). **Karim Asami:** Conceptualization (supporting); Methodology (equal); Writing – original draft (supporting); Writing – review & editing (supporting). **Claus Emmelmann:** Conceptualization (supporting); Methodology (equal); Writing – review & editing (supporting). **Ingomar Kelbassa:** Writing – original draft (supporting); Writing – review & editing (equal).

REFERENCES

[1] C. Emmelmann, M. Petersen, J. Kranz, and E. Wycisk, "Bionic Lightweight Design by Laser Additive Manufacturing (LAM) for aircraft industry," Proc. SPIE **8065**, 80650L (2011).

[2] N. F. Lepora, P. Verschure, and T. J. Prescott, "The state of the art in biomimetics," Bioinspir. Biomim. **8**, 013001 (2013).

Journal of Laser Applications ARTICLE pubs.aip.org/lia/jla

[3]C. Emmelmann, P. Sander, J. Kranz, and E. Wycisk, "Laser additive manufacturing and bionics: Redefining lightweight design," Phys. Procedia **12**, 364–368 (2011).

[4]M. Gralow, F. Weigand, D. Herzog, T. Wischeropp, and C. Emmelmann, "Biomimetic design and laser additive manufacturing—A perfect symbiosis?," J. Laser Appl. **32**, 21201 (2020).

[5]A. Du Plessis, C. Broeckhoven, I. Yadroitsava, I. Yadroitsev, C. H. Hands, R. Kunju, and D. Bhate, "Beautiful and functional: A review of biomimetic design in additive manufacturing," Addit. Manuf. **27**, 408–427 (2019).

[6]W. Nachtigall, *Bionics by Examples: 250 Scenarios from Classical to Modern Times*, 2015th ed. (Springer International, Cham, 2015).

[7]M. H. Dickinson, "Bionics: Biological insight into mechanical design," Proc. Natl. Acad. Sci. U.S.A. **96**, 14208–14209 (1999).

[8]B. Bhushan, "Biomimetics: Lessons from nature—An overview," Philos. Trans. Math. Phys. Eng. Sci. **367**, 1445–1486 (2009).

[9]T. Röver *et al.*, "Methodology for integrating biomimetic beams in abstracted topology optimization results," in *Proceedings of the ASME 2022 International Mechanical Engineering Congress and Exposition. Volume 4: Biomedical and Biotechnology; Design, Systems, and Complexity, Columbus, Ohio, 30 October–3 November* (ASME, New York, 2022).

[10]M. P. Bendsøe, M. P. Bendsøe, and O. Sigmund, *Topology Optimization: Theory, Methods and Applications* (Springer, Berlin, 2003) [Online], see http://www.loc.gov/catdir/enhancements/fy0815/2002030512-d.html

[11]COMSOL, see https://doc.comsol.com/5.4/doc/com.comsol.help.comsol/COMSOL_ReferenceManual.pdf for "Reference Manual: Version: Comsol Multiphysics 5.4 [Online]" (accessed May 25 2023).

[12]COMSOL, see https://doc.comsol.com/6.0/doc/com.comsol.help.comsol/COMSOL_ReferenceManual.pdf for "Reference Manual: Version: Comsol Multiphysics 6.0 [Online]" (accessed Mar. 31 2023).

[13]T. Röver, M. Bader, K. Asami, C. Emmelmann, and I. Kelbassa, see https://doi.org/10.15480/882.8168 for "Supplementary material to article with title: Development and assessment of a methodology for abstraction of topology optimization results to enable the substitution of optimized beams" (2023) [Online].

[14]G. Fiuk and M. W. Mrzygłód, "Numerical benchmarks for topology optimization of structures with stress constraints," Bull. Pol. Acad. Sci. Tech. Sci. **69**, e139317 (2021) [Online], see http://journals.pan.pl/Content/121172/PDF/Z_18_02049_Bpast.No.69(6)_OK.pdf

[15]O. Sigmund, "A 99 line topology optimization code written in matlab," Struct. Multidiscip. Optim. **21**, 120–127 (2001).

[16]COMSOL, see https://doc.comsol.com/5.4/doc/com.comsol.help.opt/OptimizationModuleUsersGuide.pdf for "Optimization module user's guide: Version: Comsol Multiphysics 5.4 [Online]" (accessed May 25 2023).

[17]M. P. Bendsøe, "Optimal shape design as a material distribution problem," Struct. Optim. **1**, 193–202 (1989).

[18]M. P. Bendsøe and N. Kikuchi, "Generating optimal topologies in structural design using a homogenization method," Comput. Methods Appl. Mech. Eng. **71**, 197–224 (1988).

[19]A. Nana, J.-C. Cuillière, and V. Francois, "Automatic reconstruction of beam structures from 3D topology optimization results," Comput. Struct. **189**, 62–82 (2017).

[20]J.-C. Cuillière, V. François, and A. Nana, "Automatic construction of structural CAD models from 3D topology optimization," Comput. Aided Des. Appl. **15**, 107–121 (2018).

[21]N. Ahuja and J.-H. Chuang, "Shape representation using a generalized potential field model," IEEE Trans. Pattern Anal. Mach. Intell. **19**, 169–176 (1997).

[22]N. D. Cornea, D. Silver, X. Yuan, and R. Balasubramanian, "Computing hierarchical curve-skeletons of 3D objects," Vis. Comput. **21**, 945–955 (2005).

[23]N. D. Cornea, D. Silver, and P. Min, "Curve-skeleton properties, applications, and algorithms," IEEE Trans. Vis. Comput. Graph. **13**, 530–548 (2007).

[24]J. Hicklin *et al.*, see https://math.nist.gov/javanumerics/jama/ for "JAMA: A Java Matrix Package: The MathWorks and the National Institute of Standards and Technology (NIST) (2012) [Online]" (accessed July 11 2023).

[25]M. Hermann, *Numerische Mathematik*, 2nd ed. (Oldenbourg Verlag, Munich, 2006).

[26]R. V. Mises and H. Pollaczek-Geiringer, "Praktische verfahren der gleichungsauflösung," Z. Angew. Mathematik Mechanik. **9**, 58–77 (1929).

[27]COMSOL, see https://doc.comsol.com/6.0/doc/com.comsol.help.sme/StructuralMechanicsModuleUsersGuide.pdf for "Structural mechanics module user's guide: Version: Comsol Multiphysics 6.0 [Online]" (accessed July 11 2023).

[28]COMSOL, see https://doc.comsol.com/6.0/doc/com.comsol.help.comsol/COMSOL_ApplicationBuilderManual.pdf for "Application builder reference manual: Version: Comsol Multiphysics 6.0 [Online]" (accessed July 11 2023).

A.3 Dimensioning of biomimetic beams under bending for additively manufactured structural components

The following pages feature a reprint of the journal article “Dimensioning of biomimetic beams under bending for additively manufactured structural components,” originally published in 2023. This reprint is provided under the terms of the CC BY 4.0 license.

biomimetics

MDPI

Article

Dimensioning of Biomimetic Beams under Bending for Additively Manufactured Structural Components

Tim Röver [1,*], Cedrik Fuchs [2], Karim Asami [1] and Claus Emmelmann [1]

[1] Hamburg University of Technology (TUHH), Institute of Laser and System Technologies (iLAS), Harburger Schloßstraße 28, 21079 Hamburg, Germany; karim.asami@tuhh.de (K.A.); c.emmelmann@tuhh.de (C.E.)

[2] Centerline Design GmbH, Spitzenrade 3, 24107 Quarnbek, Germany; cedrik.fuchs@centerlinedesign.de

* Correspondence: tim.roever@tuhh.de; Tel.: +49-40-42878-5010

Abstract: Additively manufactured mechanical components show great lightweight characteristics and can often be enhanced by integrating biomimetic geometrical features. This study focuses on one specific subcase, namely the substitution of solid cylindrical beams that are under bending with geometrically more complex biomimetic beams. Based on the pseudo-stem of the banana plant as a role model, six geometric beam designs were derived. Given the manufacturing constraints of the PBF-LB/M process, two abstractions were selected for detailed investigation in the main part of this study. The beam lengths were set to 100 mm. Based on parametric optimization simulations, optimal design parameters were identified for the two biomimetic abstractions for 26 different bending load cases ranging from 14 to 350 Nm. Analogous parameter optimizations were performed for a solid cylindrical beam design, which was used as a reference. The results provide detailed design solutions within the investigated intervals for biomimetic beams that can be substituted into more complex mechanical component designs with ease. The analysis provides information on which structures to use for the investigated loads. With the help of the developed numerical models, designers can easily generate biomimetic beam designs for specific bending load values.

Keywords: beam structures; biomimetics; FEM; component design; lightweight design; parameter optimization; additive manufacturing; powder bed fusion; PBF-LB/M

Citation: Röver, T.; Fuchs, C.; Asami, K.; Emmelmann, C. Dimensioning of Biomimetic Beams under Bending for Additively Manufactured Structural Components. *Biomimetics* **2024**, *9*, 214. https://doi.org/10.3390/biomimetics9040214

Academic Editor: Xiaoming Zhang

Received: 29 January 2024
Revised: 26 March 2024
Accepted: 27 March 2024
Published: 4 April 2024

1. Introduction

The European Union has set the goal of developing and using sustainable mobility systems as one of the measures to achieve climate neutrality [1]. During the production of materials such as aluminum, energy is needed, and CO_2 is emitted. One strategy to increase the sustainability of mobility systems is to reduce the amount of material used for their components through lightweight design [2]. Furthermore, lightweighting was found to be important for the reduction in fuel consumption and emissions during service, especially in applications such as aviation [3].

A study in the field of lightweight design from 2011 suggests that the utilization of biomimetic beams in structural components can improve a part's design [4]. In their study, the authors used topology optimization to develop an initial design for a structural aircraft bracket and introduced a bamboo structure into the design in a subsequent step. The authors state that due to the material change from aluminum to titanium and the novel component design, a mass reduction of 50% compared to the previously used part could be achieved [4].

In the work by [5], a biomimetic alternative based on bamboo was developed for cylindrical shells under compression or bending that are used as mechanical components. Numerical analyses suggested that the developed biomimetic structure's load-bearing efficiency (18.52×10^4 [kN kg^{-1}]) is 2.248 of that of an equal-mass, hat-stiffened cylindrical shell of the same material (8.237×10^4 [kN·kg^{-1}]) [5]. This example shows the great potential of the utilization of biomimetic beams in structural mechanical problems.

In [6,7], design methodologies for the integration of biomimetic beams into structural design concepts from topology optimization are presented. Mechanical topology optimization results often contain cylindrical-shaped beam structures. The works focus on the substitution of these structures by geometrically more complex biomimetic beams similar to the ones presented in [5]. As these biomimetic beams are assumed to be lighter while supporting the loads sufficiently, this approach is assumed to lead to a decrease in the mass of the overall component design [6].

Alternatively to the use of methodologies by [6,7], biomimetic beams can be used in the manual design process to enhance topology optimization results. It is common that experienced part designers and stress engineers are needed to develop component designs based on topology optimization results. Integration of biomimetic beams adds to the complexity of the design process.

The authors of this article speculate that the use of a tool for automated dimensioning of biomimetic beams can speed up the design process considerably by providing time-saving and efficient solutions for this design step. Therefore, costs may be saved while possibly achieving better technical solutions.

Due to the complex geometry of biomimetic beams such as those presented in [5], the additive manufacturing (AM) technique of powder bed fusion of metals by a laser beam (PBF-LB/M) was considered the manufacturing technique of choice in this work. This choice is in line with [4], in which the novel bracket was produced by PBF-LB/M. Furthermore, in [8] from 2020 titled "Biomimetic design and laser additive manufacturing—a perfect symbiosis?" the authors conclude that "[...] it is still obvious that both biomimetic design and LAM can benefit from each other".

Another conclusion that is drawn in [8] is that one limitation of the combination of additive manufacturing and biomimetics is the lack of design tools for biomimetic components. This article contributes to closing this research gap by offering a design tool for biomimetic beam designs for structural components. It is noted that detailed rather than only conceptual design solutions are provided.

This article presents numerous different biomimetic beams under bending and the maximum loads that they can support. Manufacturing restrictions for PBF-LB/M were considered in the development of the designs. Therefore, the biomimetic beams presented in this article can be easily integrated into structural PBF-LB/M component designs to improve their lightweight characteristics. Furthermore, parametric optimization models can be used for the dimensioning of biomimetic beams for specific bending load values.

2. Methodology

In the first step, a biological blueprint was identified and abstracted to obtain four biomimetic beam designs. Considering the manufacturing constraints of PBF-LB/M, two additional biomimetic designs were developed. These were selected for a more detailed investigation in this study. For these two types of abstractions, parametric designs were developed, such that the choice of specific parameters for each parametric design was necessary to obtain the final dimensions of a biomimetic beam. Manufacturing constraints of the PBF-LB/M process were considered for the parametric designs to ensure manufacturability as well as additional design assumptions. The main focus of this part of the work was the design of the cross-sectional areas of the beams. Designs were developed so that the beam cross-sections do not change along the lengths of the beams.

The parametric biomimetic beams were then optimized using a parameter optimization with regard to bending. While doing so, optimizations were carried out for numerous specific loading conditions. The optimization goal was to minimize the cross-sectional areas of the beams (a measure that is proportional to their mass) while sufficiently supporting the loads without exceeding the yield strength of the material anywhere inside the beam. For comparison to conventional beam designs, a parametric beam design with a cross-sectional area of a circle was optimized. To ensure the validity of the simulation and optimization results, a mesh convergence analysis was carried out for one of the models. Finally, selected biomimetic beam samples were additively manufactured by PBF-LB/M in an aluminum alloy to demonstrate their manufacturability.

Therefore, the methodology of this work can be summarized as follows:

1. Abstraction of a biological blueprint and generation of four beam cross-section designs
2. Development of two abstractions and parametric models considering manufacturability by PBF-LB/M
3. Parametric optimization of two biomimetic beams under various bending loads
4. Parametric optimization of a solid cylindrical beam under various bending loads as a reference
5. Conduct of a mesh convergence study
6. Additive manufacturing of exemplary biomimetic beams

3. Abstraction and Parameter Optimization

In this section, parametric biological role models for beam-like structures are reviewed. Furthermore, abstractions of a biological beam structure are presented, as well as parameter optimizations and a mesh convergence analysis.

3.1. Biological Role Models

Apart from bamboo, numerous biological structures can be considered as role models for the design of biomimetic beams. The grass stem, porcupine quill, and hedgehog spine, which have beam characteristics, have been investigated regarding elastic buckling and showed high potential for biomimicry [9]. Grass stalks and horsetails were found to have considerable bending and distortion stiffness based on interconnected ring structures that allow for a beneficial axial second moment of area [10]. In [11], biomimetic infill structures based on a bird nest, a cocoon, trees, turtle shells, and bone were investigated. This study focused on additively manufactured samples and found great potential for the designs for the three objectives of maximal supported load in compression, resilience, and strength-to-weight ratio. Various biomimetic tubular metamaterials based on two cacti species were numerically investigated with regard to their torsional properties in [12]. In [13], biomimetic cylindrical shells were investigated numerically and experimentally with regard to their energy absorption capabilities. Designs were inspired by the biological role models of turtle, horsetail, bamboo, cattail, palm, and lemon. Lemon-inspired infill achieved the best results with regard to mean crush force and specific energy absorption [13].

Considering the study at hand focuses on the bending stiffness of biomimetic beams, grass-stalks, bamboo, and horsetail were the role models found in the literature that were of the highest relevance for this study. Figure 1 shows the cut surface of the pseudo-stem of a banana plant. The structure consists of leaves placed together, which in turn have lightweight construction characteristics: massive, heavily reinforced outer skins (epidermises), held at a distance by supports. The single layers show similarities to a sandwich structure. The overall structure shows high rigidity [14]. Due to these interesting characteristics, the authors assumed a high potential for the bending stiffness of biomimetic beams based on the banana plant pseudo-stem. Therefore, the pseudo-stem of the banana plant was analyzed in more detail in this study.

3.2. Abstraction of Banana Pseudo-Stem for Additive Manufacturing

The biomimetic abstractions into parametric geometrical designs are presented in the following section. In this work, the structures of the banana pseudo-stem were used as role models for the abstractions due to their high potential for mechanical applications.

Figure 1. Photograph of the cut surface of the pseudo-stem of a banana plant. Reprinted from [15].

Figure 2a–d show the biomimetic abstractions that were derived from the pseudo-stem. Rather than having numerous cavity rings as in the biological role model, abstractions with one cavity ring and abstractions with two cavity rings were derived. This is due to the manufacturing restrictions that were considered. Considering a minimum distance between opposing walls and minimum wall thicknesses, having more than two cavity rings would lead to beams with a rather large outer diameter that was expected to be less relevant for mechanical designs. This aspect is connected to the fact that the build envelope of a commonly used PBF-LB/M system is 500 mm × 280 mm × 365 mm [16] and comparable to other machines that are used in the AM industry.

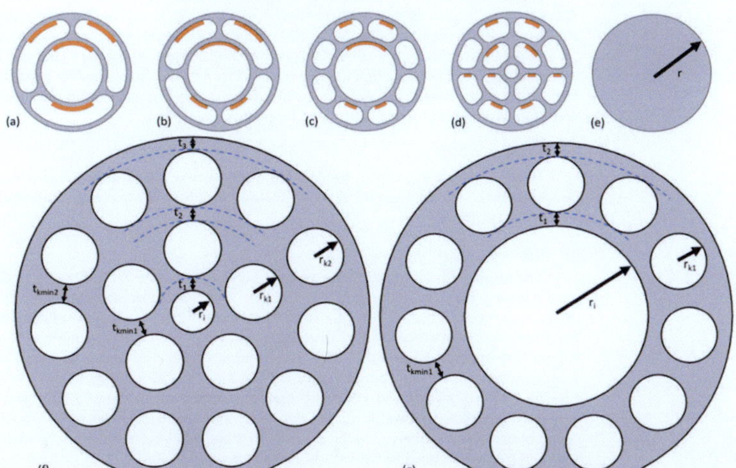

Figure 2. **(a–d)** Intermediate biomimetic abstractions of the banana plant pseudo-stem in the form of parametric cross-sectional designs of beams. Problematic areas considering overhang manufacturing constraints by PBF-LB/M are highlighted in orange; **(e)** Solid cylindrical beam; **(f,g)** Biomimetic abstractions of banana plan pseudo-stem adapted for manufacturing by PBF-LB/M, were named "revolver drum" beams.

Considering the number of cavities and the number of cavity rings, four parametric designs were derived. Three designs with one cavity ring were derived with three, four, and eight cavities, each (Figure 2a–c). For designs with one cavity ring, the one with eight cavities is the one most similar to the biological blueprint due to it having the highest number of cavities.

Furthermore, one design with two cavity rings was derived, in which the inner cavity ring consists of four cavities and the outer cavity ring consists of eight cavities (see Figure 2d).

In many AM processes, such as PBF-LB/M, support structures are necessary when parts of the design have a particular overhang angle. Support structures are a cost driver for the final part based on factors such as additional material usage, longer fabrication time, and the removal of support structures [17]. A study [18] achieved good geometrical accuracy for unsupported overhanging surfaces with an overhang angle of 45°. The samples were manufactured from AlSi10Mg [18]. Therefore, a limit overhang angle of 45° was assumed in this work. Areas that might be problematic with regard to the overhang constraint are highlighted in Figure 2a–d. Considering the main axes of the beams are parallel to the build platform (with an orientation of 0°), internal support structures would be needed for the majority of possible designs. For orientations in which the main axis is at 45–90° with respect to the build platform, no supports would be necessary within the cavities. However, it was found desirable to generate biomimetic beam designs that can be manufactured in PBF-LB/M without internal support structures, independent of their orientation in the build chamber.

Figure 2f,g show two parametric designs that were developed to generate designs that can be manufactured in any orientation by PBF-LB/M without the need for support structures within the beams (within certain intervals for each design parameter).

These structures were named "revolver drum" beams. As can be seen from Figure 2, one revolver drum beam design has one cavity ring, and the other revolver drum beam has two cavity rings. As the revolver drum structures can be manufactured more easily, they were focused on and further optimized by parameter optimization.

3.3. Comparison Structure

To evaluate the benefits of using biomimetic beams, a conventional structure with a cross-section of a solid cylindrical beam (SC) was used for comparison. The structure can be seen in Figure 2e.

3.4. Material

The aluminum alloy AlSi10Mg was chosen as the material for all parts of this work as it is widely used in lightweight applications. Materials processed by PBF-LB/M often show anisotropy. Application of a heat treatment of 2 h at 300 °C was considered for the material parameters used in this study, as a suitable heat treatment was found to reduce anisotropy [19] in AlSi10Mg. The yield strengths of horizontally and vertically manufactured AlSi10Mg samples are 141 MPa and 142 MPa, respectively. The Young's moduli of horizontally and vertically manufactured samples are 59 GPa and 57 GPa, respectively [20]. For simplification of this study and based on the fact that the anisotropy of the material is relatively low, the material in this study was assumed to be isotopic. The material properties used for this study can be found in Table 1.

Table 1. Material properties of heat-treated AlSi10Mg manufactured by PBF-LB.

Material Property	Value
Yield strength	141 MPa [20]
Young's modulus	57 GPa [20]
Poisson's ratio	0.34 [21]
Density	2.67 g/cm^3 [20]

3.5. Parameter Optimization

The parametric biomimetic beams were optimized using a parameter optimization with regard to bending. Parametric optimizations and mesh convergence analysis were carried out in COMSOL Multiphysics 6.0 [22].

Numerous specific bending loads were considered, namely 26 different values for the bending load between 14 and 350 Nm. The optimization goal was to minimize the cross-sectional areas of the beams (as a proportional measure of their mass) while sufficiently supporting the loads without exceeding the yield strength of the material anywhere inside the beam. All optimizations were carried out considering the constant lengths of all beams of l_B = 100 mm.

For this work, simplifying assumptions were made. Only linear deformations and stationary loads were considered in the parametric optimizations. The more general optimization problem as implemented in COMSOL Multiphysics 6.0 can be written as

$$\left.\begin{array}{c} \min_{\xi} Q(u(\xi),\xi) \\ L(u(\xi),\xi) = 0 \\ lb_P \leq P(u(\xi),\xi) \leq ub_P \\ lb_\Psi \leq \Psi(\xi) \leq ub_\Psi \\ lb_b \leq \xi \leq ub_b \end{array}\right\} \quad (1)$$

with control variables ξ, the scalar-valued objective function Q, the PDE solution u, and the discretized PDE ($L(u(\xi),\xi) = 0$) [23].

The constraints are divided into performance constraints on $P(u(\xi),\xi)$ that mix u and ξ, constraints on general expressions of the control variables $\Psi(\xi)$, and constraints directly on the control variables ξ [23].

The objective functions used in this work focus on minimization of the cross-sectional areas of the beam designs and can be written as

$$\left.\begin{array}{c} \min_{\xi} A(u(\xi),\xi) \\ \xi \in C \end{array}\right\} \quad (2)$$

with A being the cross-sectional area of the beam and C being the feasible set.

The summarized constraints for parameter optimization of the RD1 design that define the feasible set as used in this work can be written as:

$$\left.\begin{array}{c} 1.5\text{ mm} \leq r_{i,RD1} \leq 7.5\text{ mm} \\ 1.0\text{ mm} \leq t_{1,RD1} \leq 9.5\text{ mm} \\ 1.0\text{ mm} \leq t_{kmin,RD1} \leq 2 \times (r_{i,RD1} + t_{1,RD1}) \\ 1.5\text{ mm} \leq r_{k,RD1} \leq 7.5\text{ mm} \\ 1.0\text{ mm} \leq t_{2,RD1} \leq 9.5\text{ mm} \end{array}\right\} \quad (3)$$

and

$$\left.\begin{array}{c} r_{o,RD1} = r_{i,RD1} + t_{1,RD1} + t_{2,RD1} + 2 \times r_{k,RD1} \leq 15\text{ mm} \\ \sigma_{max,RD1} \leq 141\text{ MPa} \end{array}\right\}. \quad (4)$$

The summarized constraints for parameter optimization of the RD2 design that define the feasible set as used in this work can be written as:

$$\left.\begin{array}{c} 1.5\text{ mm} \leq r_{i,RD2} \leq 7.5\text{ mm} \\ 1.0\text{ mm} \leq t_{1,RD2} \leq 5.5\text{ mm} \\ 1.0\text{ mm} \leq t_{kmin1,RD2} \leq 2 \times (r_{i,RD2} + t_{1,RD2}) \\ 1.5\text{ mm} \leq r_{k1,RD2} \leq 7.5\text{ mm} \\ 1.0\text{ mm} \leq t_{2,RD2} \leq 5.5\text{ mm} \\ 1.0\text{ mm} \leq t_{kmin2,RD2} \leq 2 \times (r_{i,RD2} + t_{1,RD2} + 2 \times r_{k1,RD2} + t_{2,RD2}) \\ 1.5\text{ mm} \leq r_{k2,RD2} \leq 7.5\text{ mm} \\ 1.0\text{ mm} \leq t_{3,RD2} \leq 5.5\text{ mm} \end{array}\right\} \quad (5)$$

and

$$\left.\begin{array}{r} r_{o,RD2} = r_{i,RD2} + t_{1,RD2} + t_{2,RD2} + t_{3,RD2} + 2 \times r_{k1,RD2} + 2 \times r_{k2,RD2} \leq 15\text{ mm} \\ \sigma_{max,RD2} \leq 141\text{ MPa} \end{array}\right\}. \quad (6)$$

The summarized constraints for parameter optimization of the SC design that define the feasible set as used in this work can be written as:

$$1.5\text{ mm} \leq r_{SC} \leq 15\text{ mm} \quad (7)$$

and

$$\left.\begin{array}{r} r_{SC} \leq 15\text{ mm} \\ \sigma_{max,SC} \leq 141\text{ MPa} \end{array}\right\} \quad (8)$$

The constraints were developed based on the following: It was assumed that the radius of any of the cavities had to be equal to or larger than 1.5 mm to ensure powder removability in the PBF-LB process. At the same time, the radius of any cavity has to be equal to or smaller than 7.5 mm to ensure support-free manufacturability based on [24]. Furthermore, the minimum allowable wall thickness was defined as 1 mm to ensure manufacturability and additionally reduce the chances of local buckling in thin walls.

The upper limits for occurring von Mises stresses in the domains of interest are given by the yield strength of AlSi10Mg manufactured by PBF-LB, which is 141 MPa based on [20].

The constraints on the outer radii of the beams in the optimization were defined to be 15 mm. Therefore, the ratio between the length of the beams in this study (100 mm) and their maximum diameter (30 mm) would ensure a beam-like character of the design. Certain upper bounds of Equations (3) and (5) represent the maximum value for the respective parameter, considering that the remaining design parameters are set to their lower bounds while respecting a maximum diameter of the beam of 30 mm. By doing so, the solution space was further restricted to promote faster convergence of the optimization.

For parametric optimization, the popular Nelder–Mead method [25,26] as implemented in COMSOL Multiphysics 6.0 [23] was used.

The constraints according to Equations (4) and (6) were implemented using an augmented Lagrangian method based on a function implemented in COMSOL Multiphysics 6.0 based on the theory of Lagrange multipliers [23].

Parameter optimizations were conducted to find the optimal combination of design parameters for each load case, based on the information previously described. It is noted that the Nelder–Mead solver may have identified local minima of the global minimization problem. Therefore, the presented results may not be the global optimal solutions to the problems. This, however, is a common problem in non-linear optimization [27].

During each optimization simulation, multiple finite-element models were solved iteratively (based on the variation in design parameters) using automatically generated meshes. The displacement fields of all elements were chosen as quadratic serendipity to achieve results of high quality. More information regarding finite element meshes can be found in Section 3.5.

Figure 3 shows an exemplary RD1 design. The total lengths of the beam structures in the optimizations were 101 mm. However, the relevant lengths of the beams that were investigated were 100 mm. In the following, this matter is explained in more detail, and the boundary conditions and evaluation domains are also presented.

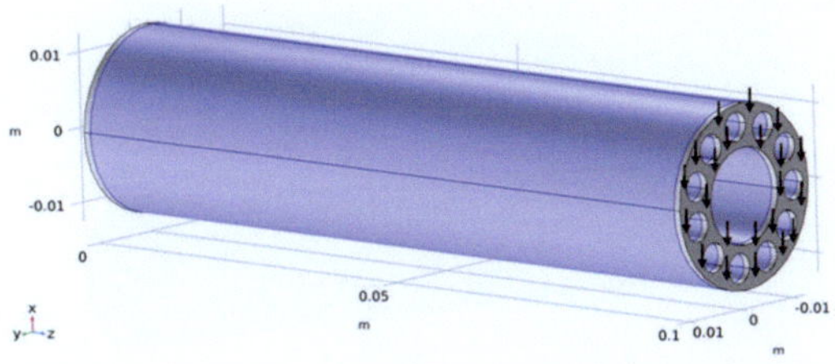

Figure 3. Exemplary optimization model with volume considered for stress evaluation highlighted. The model was mechanically fixed at z = 0 m, and a boundary force was applied to the free end.

The models were divided into three domains:

- Domain 1: 0 mm $\leq$ z < 1 mm
- Domain 2: 1 mm $\leq$ z $\leq$ 100 mm
- Domain 3: 100 mm < z $\leq$ 101 mm

The beams were mechanically fixed at one end by fixing the respective circular end surface. For the application of a bending moment, the end surface of the beam on its other side was subjected to a shear force in the negative x-direction (orthogonal to the main axis of the beams).

Von Mises stresses (see Equations (4) and (6)) in the models were evaluated in domain 2. The respective volume for which von Mises stresses were evaluated is highlighted in Figure 3. Consequently, stresses were not evaluated close to the perfectly stiff fixations as well as close to the surfaces to which the force was applied. This prevents the physically meaningless evaluation of stress peaks that are due to the discretization of the mechanical problem.

At the same time, the highest stresses of interest that are expected to appear in the regions at z = 1 mm are 100 mm from the surface of the application of force. Therefore, the relevant length of the investigated beam is 100 mm. The applied forces at the free ends were varied, with values of forces ranging from 140 N to 3500 N, corresponding to 14 Nm and 350 Nm bending loads, respectively. Therefore, 26 variations for each of the parametric designs (RD1, RD2, and SC) were calculated. Based on the lever (100 mm), the maximum bending moments for each model inside the evaluated volume are given in the tables in Appendix A.

3.6. Discretization

To ensure low mesh-dependency of the optimization results, a mesh convergence analysis of an exemplary design was carried out. All meshes were developed using the same methodology and functions as implemented in COMSOL Multiphysics 6.0 [22]. The meshing methodology as well as the mesh convergence study are presented below. Figure 4 shows the finite-element mesh of an exemplary RD2 model. The development of the meshes for both parametric optimizations was conducted by:

1. Automatic meshing of the end surface using triangular elements (highlighted surface in Figure 4) based on five parameters for automatic meshing. The parameters were maximum element size, minimum element size, maximum element growth rate, curvature factor, and resolution of narrow regions.
2. Development of prism elements based on triangular mesh on the surface and the swept function. The thickness of the prism elements in domains 1 and 3 was chosen to be 0.25 mm (4 layers). For the rest of the beam, the thickness of the prisms was chosen to be 3 mm (33 layers).

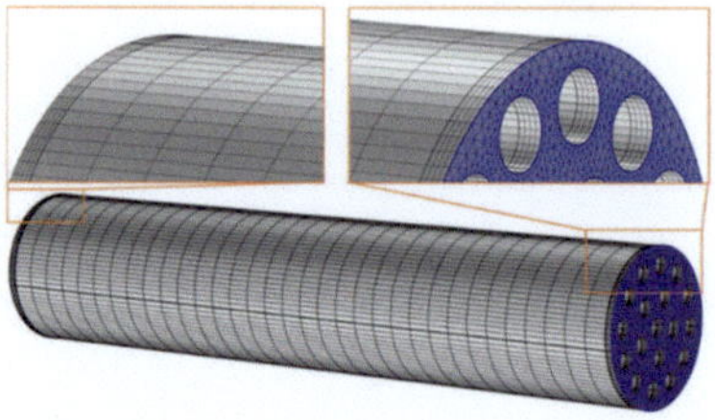

Figure 4. Finite-Element mesh of an exemplary RD2 model.

The values of the previously mentioned meshing parameters can be found in Table 2. The values for maximum element growth rate, curvature factor, and resolution of narrow regions were defaulted in the software and were not changed. The other parameters were chosen based on a mesh convergence study, which is briefly presented below.

Table 2. Meshing parameters.

Meshing Parameter	Value
Maximum element size ($S_{el,max}$)	2×10^{-2} m
Minimum element size ($S_{el,min}$)	2×10^{-4} m
Maximum element growth rate	1.3
Curvature factor	0.2
Resolution of narrow regions	1.0
Number of prism element layers in domains 1 and 3 ($N_{PL,D1,3}$)	4
Number of prism element layers in domain 2 ($N_{PL,D2}$)	33

3.7. Mesh Convergence Study

The mesh convergence study was conducted with the RD2 design, as it was assumed to be the most critical one with regard to mesh dependency based on its geometric complexity. A RD2 design with the smallest possible design parameters and a downward force of 500 N at the free end was used. The parameters $S_{el,max}$ and $S_{el,min}$ were varied using multiplication factors of 0.5, 1, and 2. Analogously, the parameter $N_{PL,D1,3}$ was varied using multiplication factors of 0.5, 1, and 2. Furthermore, the parameter $N_{PL,D2}$ was varied using multiplication factors of 1/3, 1, and 3. Considering all possible combinations of these three variations, a total of 18 combinations of meshing parameters were analyzed.

The model of the mesh convergence study with the most elements ($S_{el,max} = 0.5 \times 2 \times 10^{-2}$ m; $S_{el,min} = 0.5 \times 2 \times 10^{-4}$ m; $N_{PL,D1,3} = 2 \times 4$; $N_{PL,D2} = 3 \times 33$) was considered the reference model for the other combinations, as it was assumed to be the most accurate. Its total displacement at the free end was evaluated as 4.77×10^{-4} m, and a maximum stress in domain 2 of 79.2 MPa was evaluated. The maximum absolute relative error of the displacement at the free end of all 17 models in this study with respect to the reference model was 0.069%. The maximum absolute relative error of the maximum stress in domain 2 of all 17 models in this study with respect to the reference model was 6.394%. The model with medium multiplication factors ($S_{el,max} = 1 \times 2 \times 10^{-2}$ m; $S_{el,min} = 1 \times 2 \times 10^{-4}$ m; $N_{PL,D1,3} = 1 \times 4$; $N_{PL,D2} = 1 \times 33$) showed relative errors for the displacement at the free end and that of the maximum stress in domain 2 of 0.006% and 2.235%, respectively.

Considering the influence of the number of elements in the finite element mesh on the computation time and the accuracy of the results of the numerical model, a balance had to be found. From the mesh convergence study, it was concluded that medium multiplication factors were a good choice for the parameter optimizations and were, therefore, used in this work. Research data from the mesh convergence study was made available in the Supplementary Materials.

4. Results

In this section, the numerical results of the parameter optimization and their evaluation are presented. Additionally, results regarding additive manufacturing of exemplary designs of RD1 and RD2 are presented.

4.1. Numerical Results

For each parametric design and investigated bending force value, one set of design parameters was identified as optimal by parametric optimization using the developed models. These optimal sets of geometric parameters for each case are listed in the tables in Appendix A. It could be confirmed that the occurring stresses did not exceed the material's yield strength in any of the numerically investigated models. For an efficient overall component design, it has to be decided whether RD1, RD2, or SC should be used to substitute a beam. Figure 5 summarizes the results of the parameter optimization. The graph shows the cross-sectional areas of the beams over the applied bending forces. Considering lightweight engineering, a low cross-sectional area is beneficial.

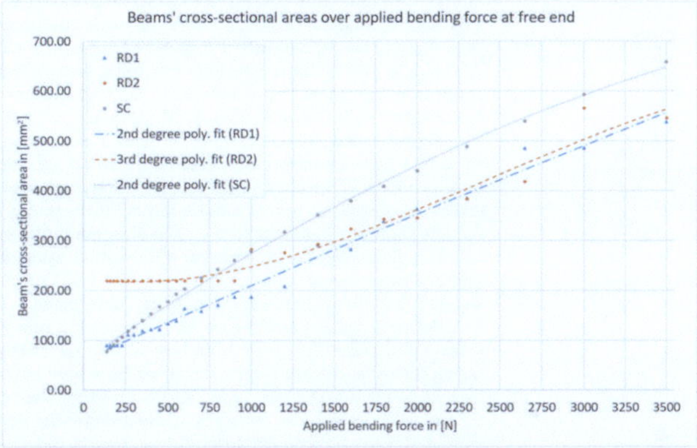

Figure 5. Results of the numerical parameter optimization study: Cross-sectional area over applied bending force at the free end for revolver drum beam 1 (RD1, blue), revolver drum beam 2 (RD2, orange), and solid cylindrical beam (SC, grey).

All three curves have an upward trend, as expected. The curves of biomimetic beams RD1 and RD2 have a plateau for low bending force values. This is due to the fact that the minimum possible dimensions of the designs successfully support numerous of the small-valued load cases. This effect is stronger for RD2 than for RD1. It can be seen that for low bending forces up to 160 N (16 Nm), an SC beam is favorable. For bending forces between 180 N (18 Nm) and 1800 N (180 Nm), RD1 is favorable. Between 2000 N (200 Nm) and 3500 N (350 Nm), the data does not allow for a general statement on whether RD1 or RD2 is more favorable. However, the preferable design between the two can be selected for a certain load case based on Figure 5.

Therefore, using Figure 5 and the tables in Appendix A of this article, an optimized beam for component design can be realized with little effort for the investigated load cases.

Polynomial fits in Figure 5 support the previous statements that for low force values, SC beams are favorable, for medium force values, RD1 beams are beneficial, and for high force values, optimized RD1 and RD2 beams should be compared to identify the lighter beam.

TIn the following, characteristics of designs RD1, RD2, and SC are discussed: In general, regarding bending loads, the distance of material from the neutral axis in the cross-section of a beam is important. The relevant measure is the axial second moment of area [28]. Therefore, the solid cylinder is a rather disadvantageous design for bending loads since most of the material is located close to the neutral axis. Comparing designs RD1 and RD2 to the SC design, it is beneficial that RD1 and RD2 have cavities at their centers as well as additional cavity rings such that considerable amounts of mass are moved away from the neutral axis, thereby increasing the axial second moment of area.

Based on the geometrical constraints that were defined, the SC design allows for the smallest diameter of the three designs as well as the smallest cross-sectional area and smallest axial second moment of area. At the same time, it allows for the largest cross-sectional area and the largest axial second moment of area at the maximum diameter defined for the designs. However, the lightweight design character of the design with respect to bending loads is rather low. Designs RD1 and RD2 are more strongly constrained. The intervals for minimum and maximum cross-sectional area and minimum and maximum axial second moment of area, indirectly given by the geometrical constraints, are shorter than those of the SC parametric design. At the same time, it was expected that RD1 and RD2 would give better results under bending than SC beams. The results (see Figure 5) confirm that for most load values, RD1 and RD2 designs indeed provide solutions with better lightweight character than the SC designs.

Figure 6 shows the von Mises stresses in six exemplary optimization results. The relevant design parameters were taken from the optimization results, as stated in the tables in Appendix A. It can be seen that for all models, high stresses occur close to the supported ends, as was expected based on beam theory. Furthermore, it can be seen that all exemplary models show stresses close to the yield strength of the material. This is beneficial with regard to lightweight engineering and the optimization objective. If the maximum stress within a model were considerably below the yield strength of the material, this would suggest a rather low lightweight character of the beam.

Both SC designs (see Figure 6a,d) are well in line with expectations. The highest stresses occur near the supported ends of the beams (at a large distance from the downward-facing forces) and at the largest distances from the neutral axes of the beams. Both RD1 designs (see Figure 6b,e) give good results regarding their cross-sectional area compared to SC and RD2 designs (see Figure 5), while it is striking that the ratio between parameters $r_{i,RD1}$ and $r_{k,RD1}$ is considerable for the two results. Based on these unintuitive designs and parameters as optimization results, the authors conclude that parametric optimization can help product designers generate efficient and novel beam designs for unique load cases. With regard to the exemplary RD2 designs shown in Figure 6c,f, it can be seen that they also exhibit strong differences. The large inner cavity of RD2-2300 N in Figure 6f helps to move material away from the neutral axis and achieve a great axial second moment of area. For RD2-1200 N in Figure 6c, it can be seen that the value for $t_{kmin2,RD2}$ results in a considerable distance between the cavities of cavity ring 2 of the design. This results in bulk material without cavities at the outer regions of the cross-section and a beneficial axial second moment of area. Also, in this case, the qualitatively different results show that parameter optimization can provide unintuitive but efficient design solutions.

Further qualitative analysis of the optimization results revealed that the majority of optimized beams for RD1 and RD2 had relatively small radii for the cavities of the cavity rings and relatively low distances between the cavities of the cavity rings. Furthermore, the majority of beams had a relatively large radius for the center cavity. At the same time, numerous optimized beams differed qualitatively from this trend, as indicated by the results shown in Figure 6. For RD1, a trend could be identified of increasing radii of the cavities of the cavity ring with increasing bending loads (1400–3500 N). No qualitative design trends could be identified for the results of RD2.

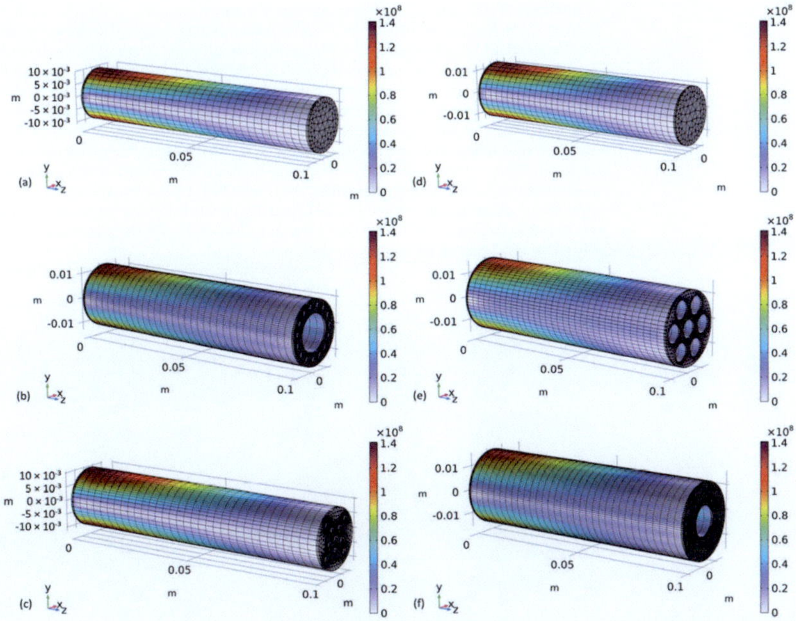

Figure 6. Von Mises stresses in exemplary optimization results. (**a**) SC optimized for 1200 N; (**b**) RD1 optimized for 1200 N; (**c**) RD2 optimized for 1200 N; (**d**) SC optimized for 2300 N; (**e**) RD1 optimized for 2300 N; (**f**) RD2 optimized for 2300 N.

Optimization simulation files of each investigated type of beam, covering each investigated type of loading, were made available in the Supplementary Materials of this article. Using these files, unique values for the bending load can be given as an input to the parameter optimizations. Furthermore, the lengths of the beams can be adapted to finally obtain customized, optimized parameters for one of the biomimetic beams. However, when the length is adapted, the number of layers of prism elements in domain 2 should be changed proportionally. The authors see great potential for the use of the provided models for systematic and reproducible biomimetic design of mechanical components in research and industry.

4.2. *Additive Manufacturing*

Figure 7 shows additively manufactured revolver drum beams. Exemplary versions of parametric designs RD1 and RD2 were produced in orientations of 0°, 45°, and 90°. Data preparation for additive manufacturing was performed with Autodesk Netfabb Ultimate slicer software (version 2023.1) [29]. Production in aluminum alloy AlSi10Mg was carried out using the PBF-LB/M system MPrint from One Click Metal [30]. The fabrication was performed with a laser power of 170 W, a scan speed of 1200 mm/s, a hatch distance of 0.1 mm, a layer thickness of 20 µm, and a focus diameter of 70 µm. It could be shown that the exemplary beam designs could be manufactured successfully in three different orientations (0°, 45°, and 90°) without the use of support structures within the cylindrical cavities of the beams. In Figure 7b, it can be seen that the internal channels of the 0°-RD2 beam could be manufactured without internal supports in the cavities. It could be confirmed that all other manufactured beams of the build job could also be manufactured without internal supports. Visual inspection revealed no defects or geometric irregularities (see Figure 7a–c). Therefore, it can be expected that designs RD1 and RD2 are well-suited for substitution in mechanical components in terms of manufacturing constraints in the PBF-LB/M process when using AlSi10Mg as a material. It is noted that powder removability has to be ensured for the powder within the cavities of the beams. With regard to the mechanical properties of the manufactured beams by PBF-LB/M, the influence of different orientations in the build volume has to be briefly considered. In that respect, it can be assumed that their orientation only has a minor influence, as relevant material parameters only show low levels of anisotropy, as previously stated.

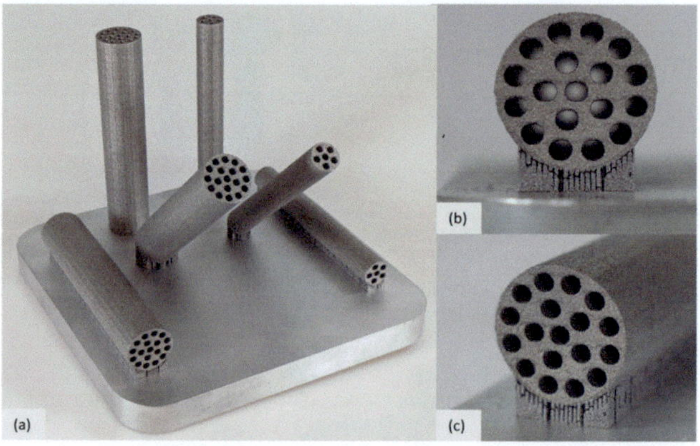

Figure 7. Exemplary revolver drum beams (RD1 and RD2) in three different orientations (0°, 45°, and 90°) manufactured by PBF LB/M in AlSi10Mg. The build platform has a base area of 152 mm × 152 mm. (**a**) Photograph of the entire build job; (**b**) Close-up of the RD2 design manufactured at 0°, showing that cavities are free of support structures; (**c**) Additional close-up of the RD2 design manufactured at 0°, showing no visible defects or geometric irregularities.

5. Conclusions and Future Prospects

Additively manufactured mechanical components show great lightweight characteristics. Such component designs can often be improved by the integration of biomimetic geometric features. This work focuses on one specific subcase, namely the substitution of solid cylindrical beams that are under bending by geometrically more complex biomimetic beams.

Based on the pseudo-stem of the banana plant as a role model, six geometric beam designs were derived. Considering the manufacturing constraints of the PBF-LB/M process, two abstractions were chosen for detailed investigation in the main part of this study. Beam lengths were chosen to be 100 mm.

Based on parametric optimization simulations, optimal design parameters were identified for the two biomimetic abstractions for 26 different bending load cases with values between 14 and 350 Nm. Furthermore, analogous parameter optimizations were performed for a solid cylindrical beam design that was used as a reference.

The results give detailed design solutions within the investigated intervals for biomimetic beams that can be substituted into more complex mechanical component designs with ease.

The analysis shows that SC beams are beneficial for small bending loads, RD1 beams are beneficial for medium values, and for higher bending loads, both RD1 and RD2 designs should be considered to choose the most beneficial design. Furthermore, the results show that, in some cases, rather unintuitive values for the design parameters were obtained by the parametric optimizations. This suggests that the provided methods and optimization models have a high potential for the design of biomimetic mechanical components.

The authors made the optimization models available in the Supplementary Materials. With the help of these models, designers can easily generate optimized beam designs for SC, RD1, and RD2 for specific bending load values.

Exemplary beam designs were produced in aluminum alloy AlSi10Mg by the additive manufacturing process PBF-LB/M. It could be shown, that the designs could be manufactured successfully without support structures inside the cavities. Therefore, the designs are well-suited for use in lightweight components manufactured by additive manufacturing.

The authors plan to extend the optimization simulations so that combined load cases considering bending moments, normal forces, shear forces, and torsion forces can be considered within one model. It is noted that the way the models were set up with one fixed end and one end on which the bending force acts is beneficial for the advancement that is planned for the optimization models. Furthermore, parameter optimizations should be developed for the remaining four abstractions presented in this study.

Furthermore, additional parametric beam designs should be developed analogously to allow for a greater solution space. Common beam designs such as I-beams and circular tubes should be investigated in future works, as well as the development of parametric beam designs based on 3D topology optimization results. Additionally, other biological blueprints, such as grass stalks, bamboo, horsetail, and the banana plant petiole, could be used for the development of parametric beam designs. The banana plant petiole could be of special interest because of the inner structures that are at approximately 45° and, therefore, well in line with overhang manufacturing constraints in the PBF-LB/M process (see Figure 8).

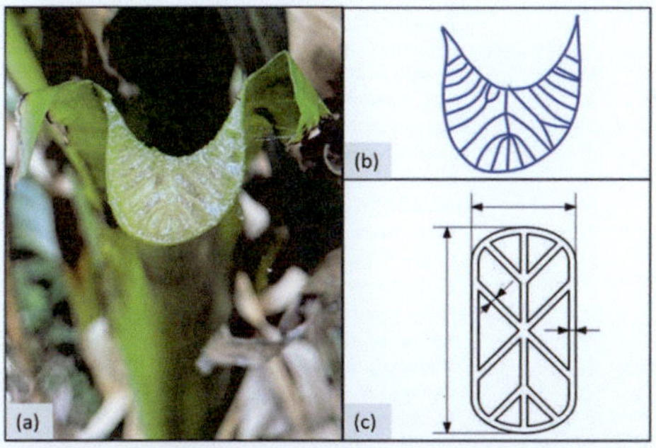

Figure 8. Exemplary abstraction of banana plant petiole cut surface: (**a**): Photograph; (**b**): Contour line; (**c**): Abstraction.

Supplementary Materials: The following supporting information [31] can be downloaded at: https://doi.org/10.15480/882.9113, Portable Document Format file S0: Readme of dataset (file name: 0_Readme.pdf); COMSOL Multiphysics 6.0 file S1: Parameter optimization of RD1 under bending by 140–700 N (file name: 1_RD1_bending_140–700.mph); COMSOL Multiphysics 6.0 file

S2: Parameter optimization of RD1 under bending by 800–3500 N (file name: 2_RD1_bending_800–3500.mph); COMSOL Multiphysics 6.0 file S3: Parameter optimization of RD2 under bending by 140–700 N (file name: 3_RD2_bending_140–700.mph); COMSOL Multiphysics 6.0 file S4: Parameter optimization of RD2 under bending by 800–3500 N (file name: 4_RD2_bending_800–3500.mph); COMSOL Multiphysics 6.0 file S5: Parameter optimization of SC under bending by 140–3500 N (file name: 5_SC_bending_140–3500.mph); COMSOL Multiphysics 6.0 file S6: Mesh Convergence Study (file name: 6_RD2_mesh_con_parametric_sweep.mph); Portable Document Format file S7: Mesh convergence evaluation overview (file name: 7_mesh_con_evaluation_overview.pdf).

Author Contributions: Conceptualization: T.R., C.F. and C.E.; methodology: T.R. and C.F.; validation: T.R. and C.F.; formal analysis: T.R.; investigation: T.R.; writing—original draft: T.R.; writing—review and editing: T.R. and K.A.; visualization: T.R. and C.F.; supervision: C.E. All authors have read and agreed to the published version of the manuscript.

Funding: Publishing fees are supported by the Funding Programme Open Access Publishing of Hamburg University of Technology (TUHH). Apart from this, this research received no external funding.

Institutional Review Board Statement: Not applicable.

Data Availability Statement: Data are contained within the article and the Supplementary Materials of this article.

Acknowledgments: The authors thank Fritz Lange (3D Spark GmbH) for valuable discussions throughout this work as well as Arthur Seibel (Leuphana University Lüneburg) and Katharina Bartsch (Fraunhofer IAPT) for valuable remarks about the written manuscript. Furthermore, the authors want to thank Jürgen Wittkamp (TUHH) for the additive manufacturing of the demonstrator part. Additionally, the authors thank the anonymous reviewers of the manuscript for their valuable comments.

Conflicts of Interest: Author C.F. was employed by the company Centerline Design GmbH. The remaining authors declare that the research was conducted in the absence of any commercial or financial relationships that could be construed as a potential conflict of interest.

Nomenclature

A	Cross-sectional area of beam
C	Feasible set
$L(u(\xi),\xi) = 0$	System of discretized partial differential equations
l_B	Length of beams (constant), 100 mm
lb_b	Lower bound of control variables
lb_P	Lower bound of performance constraint
lb_Ψ	Lower bound of general expressions of the control variables
$N_{PL,D1,3}$	Number of prism element layers in domains 1 and 3 (meshing)
$N_{PL,D2}$	Number of prism element layers in domain 2 (meshing)
$P(u(\xi),\xi)$	Performance constraints
Q	Scalar-valued objective function
r	Radius
r_i	Inner radius
r_{k1}	Radius of cavities in first cavity ring
r_{k2}	Radius of cavities in second cavity ring
r_o	Outer radius
$S_{el,max}$	Maximum element size (meshing)
$S_{el,min}$	Minimum element size (meshing)
t_{kmin1}	Minimum distance between cavities in first cavity ring
t_{kmin2}	Minimum distance between cavities in second cavity ring
t_1	Wall thickness of first ring
t_2	Wall thickness of second ring
t_3	Wall thickness of third ring
u	PDE solution
ub_b	Lower bound of control variables
ub_P	Upper bound of performance constraint

ub_{Ψ}	Upper bound of general expressions of the control variables
x_{P1}	Arbitrary design parameter P_1
x_{P2}	Arbitrary design parameter P_2
x_{Pn}	Arbitrary design parameter P_N
ξ	Control variables
σ_{max}	Maximum allowable von Mises stress
$\Psi(\xi)$	General expressions of the control variables
$\cdots_{RD1}$	Subscript indicating reference to revolver drum beam 1
$\cdots_{RD2}$	Subscript indicating reference to revolver drum beam 2
$\cdots_{SC}$	Subscript indicating reference to solid cylindrical beam

Appendix A

Table A1. Parameter optimization results for revolver drum 1 under bending loads.

Revolver drum 1 (RD1)

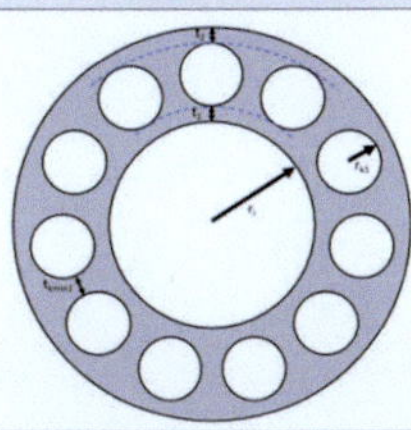

Model parameters			
Young's modulus	Poisson's ratio	Density	Length
57 [GPa]	0.34	2.67 [g/cm^3]	100 [mm]
Design variables			
Parameter	Variable	Minimum	Maximum
Inner radius	r_i	1.5 [mm]	7.5 [mm]
Wall thickness first ring	t_1	1 [mm]	9.5 [mm]
Wall thickness second ring	t_2	1 [mm]	9.5 [mm]
Radius of cavities in cavity ring	r_{k1}	1.5 [mm]	7.5 [mm]
Minimum distance between cavities in cavity ring	t_{kmin1}	1.0 [mm]	$2 \times (r_i + t_1)$
Constraints			
Physical constraint	Numerical constraint	Minimum	Maximum
Yield strength of AlSi10Mg	Maximum von Mises stress in domain 2	-	141 [MPa]
Maximum diameter of 30 mm of beam	$r_i + t_1 + t_2 + 2 \times r_{k1}$	-	15 [mm]

Bending							
Force [N]	Max. Bending Moment [Nm]	r_i [mm]	t_1 [mm]	t_2 [mm]	r_{k1} [mm]	t_{kmin1} [mm]	Area [mm^2]
140	14	1.5000	1.0000	1.0000	1.5000	1.0000	90.32
160	16	1.5000	1.0000	1.0000	1.5000	1.0000	90.32
180	18	1.5000	1.0000	1.0000	1.5000	1.0000	90.32
200	20	1.5000	1.0000	1.0000	1.5000	1.0000	90.32
230	23	1.5000	1.0000	1.0000	1.5000	1.0000	90.32
265	26.5	1.5000	1.0000	1.0000	1.5000	5.0000	111.53
300	30	1.5000	1.0000	1.0000	1.5000	5.0000	111.53
350	35	3.0936	1.0000	1.0000	1.5000	1.0000	119.18
400	40	3.1825	1.0000	1.0000	1.5000	1.0000	121.97
450	45	3.1825	1.0000	1.0000	1.5000	1.0000	121.97
500	50	3.7825	1.0000	1.0000	1.5000	1.0000	133.75
550	55	4.1825	1.0000	1.0000	1.5000	1.0000	139.25
600	60	5.1825	1.0000	1.0000	1.5000	1.0000	163.60
700	70	5.0255	1.0000	1.0000	1.5000	1.0000	158.66
800	80	5.6387	1.0000	1.0000	1.5000	1.0000	170.86
900	90	6.3787	1.0000	1.0000	1.5000	1.0000	187.04
1000	100	6.3784	1.0000	1.0000	1.5000	1.0000	187.03
1200	120	7.5000	1.0000	1.0000	1.5000	1.0000	208.13
1400	140	1.8643	3.4934	1.2866	2.9185	1.2589	291.18
1600	160	6.4023	1.2577	1.2781	1.9665	2.1404	282.34
1800	180	4.2807	2.0517	2.0384	2.2948	1.3703	337.77
2000	200	4.0351	2.1814	2.1598	2.7252	1.1168	362.80
2300	230	3.7896	1.9188	1.5913	3.9043	1.7545	384.65
2650	265	2.8336	2.9109	2.2365	3.0961	4.4002	485.41
3000	300	1.7899	2.1242	3.5810	3.8958	1.0000	485.67
3500	350	1.5000	1.0000	1.0000	7.5000	1.0000	537.99

Table A2. Parameter optimization results for revolver drum 2 under bending loads.

Revolver drum 2 (RD2)

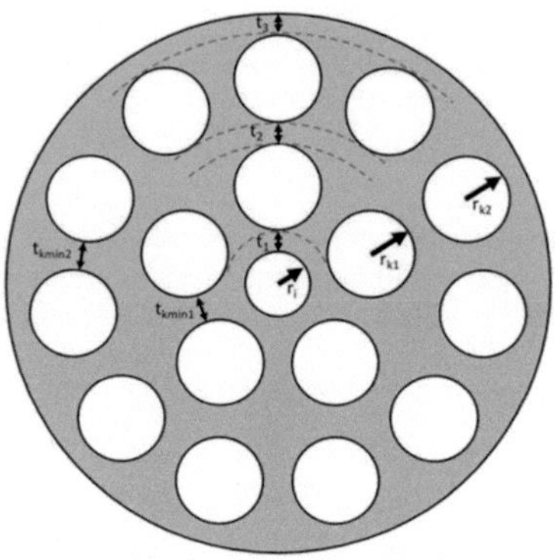

Model parameters			
Young's modulus	Poisson's ratio	Density	Length
57 [GPa]	0.34	2.67 [g/cm^3]	100 [mm]
Design variables			
Parameter	Variable	Minimum	Maximum
Inner radius	r_i	1.5 [mm]	7.5 [mm]
Wall thickness first ring	t_1	1 [mm]	5.5 [mm]
Wall thickness second ring	t_2	1 [mm]	5.5 [mm]
Wall thickness third ring	t_3	1 [mm]	5.5 [mm]
Radius of cavities in first cavity ring	r_{k1}	1.5 [mm]	7.5 [mm]
Radius of cavities in second cavity ring	r_{k2}	1.5 [mm]	7.5 [mm]
Minimum distance between cavities in first cavity ring	t_{kmin1}	1 [mm]	$2 \times (r_i + t_1)$
Minimum distance between cavities in second cavity ring	t_{kmin2}	1 [mm]	$2 \times (r_i + t_1 + 2 \times r_{k1} + t_2)$
Constraints			
Physical constraint	Numerical constraint	Minimum	Maximum
Yield strength of AlSi10Mg	Maximum von Mises stress in domain 2	-	141 [MPa]
Maximum diameter of 30 mm of beam	$r_i + t_1 + t_2 + t_3 + 2 \times r_{k1} + 2 \times r_{k2}$	-	15 [mm]

Bending										
Force [N]	Max. Bending Moment [Nm]	r_i [mm]	t_1 [mm]	t_2 [mm]	t_3 [mm]	r_{k1} [mm]	r_{k2} [mm]	t_{kmin1} [mm]	t_{kmin2} [mm]	Area [mm^2]
140	14	1.5000	1.0000	1.0000	1.0000	1.5000	1.5000	1.0000	1.0000	219.13
160	16	1.5000	1.0000	1.0000	1.0000	1.5000	1.5000	1.0000	1.0000	219.13
180	18	1.5000	1.0000	1.0000	1.0000	1.5000	1.5000	1.0000	1.0000	219.13
200	20	1.5000	1.0000	1.0000	1.0000	1.5000	1.5000	1.0000	1.0000	219.13
230	23	1.5000	1.0000	1.0000	1.0000	1.5000	1.5000	1.0000	1.0000	219.13
265	26.5	1.5000	1.0000	1.0000	1.0000	1.5000	1.5000	1.0000	1.0000	219.13
300	30	1.5000	1.0000	1.0000	1.0000	1.5000	1.5000	1.0000	1.0000	219.13
350	35	1.5000	1.0000	1.0000	1.0000	1.5000	1.5000	1.0000	1.0000	219.13
400	40	1.5000	1.0000	1.0000	1.0000	1.5000	1.5000	1.0000	1.0000	219.13
450	45	1.5000	1.0000	1.0000	1.0000	1.5000	1.5000	1.0000	1.0000	219.13
500	50	1.5000	1.0000	1.0000	1.0000	1.5000	1.5000	1.0000	1.0000	219.13
550	55	1.5000	1.0000	1.0000	1.0000	1.5000	1.5000	1.0000	1.0000	219.13
600	60	1.5000	1.0000	1.0000	1.0000	1.5000	1.5000	1.0000	1.0000	219.13
700	70	1.5000	1.0000	1.0000	1.0000	1.5000	1.5000	1.0000	1.0000	219.13
800	80	1.5000	1.0000	1.0000	1.0000	1.5000	1.5000	1.0000	1.0000	219.13
900	90	1.5000	1.0000	1.0000	1.0000	1.5000	1.5000	1.0000	1.0000	219.13
1000	100	1.6138	1.0000	1.0000	1.0000	1.5000	1.5000	1.0000	10.7700	282.11
1200	120	1.5000	1.0000	1.0000	1.0000	1.5000	1.5000	1.1546	7.0727	275.67
1400	140	1.8967	1.0353	1.0000	1.4824	1.6326	1.5815	1.0182	1.5059	292.63
1600	160	3.3163	1.1083	1.0000	1.0000	1.5000	1.5000	1.3440	2.9104	323.18
1800	180	5.1825	1.0000	1.0000	1.0000	1.5000	1.5000	1.0000	1.0000	342.54
2000	200	5.3522	1.0000	1.0000	1.0000	1.5000	1.5000	1.0000	1.0000	345.07
2300	230	6.4000	1.0000	1.0000	1.0000	1.5000	1.5000	1.0000	1.0000	383.12
2650	265	7.4000	1.0000	1.0000	1.0000	1.5000	1.5000	1.0000	1.0000	418.46
3000	300	1.5250	1.0000	1.0000	5.4000	1.5000	1.5000	1.0083	1.0083	565.26
3500	350	5.3117	1.6190	1.0019	1.3417	1.8887	1.9413	1.7177	2.0912	545.67

Biomimetics **2024**, *9*, 214

Table A3. Parameter optimization results for the solid cylindrical beam under bending loads.

Solid cylindrical beam (SC)	Model parameters			
	Young's modulus	Poisson's ratio	Density	Length
	57 [GPa]	0.34	2.67 [g/cm^3]	100 [mm]
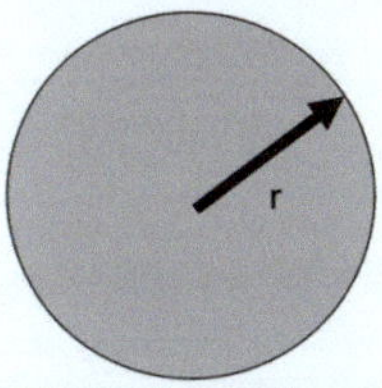	Design variables			
	Parameter	Variable	Minimur	Maximum
	Inner radius	r	1.5 [mm]	15 [mm]
	Constraints			
	Physical constraint	Numerical constraint	Minimur	Maximum
	Yield strength of AlSi10Mg	Maximum von Mises stress in domain 2	-	141 [MPa]
	Maximum diameter of 30 mm of beam	r	-	15 [mm]

Bending			
Force [N]	Max. Bending Moment [Nm]	r [mm]	Area [mm^2]
140	14	4.9805	77.93
160	16	5.1914	84.67
180	18	5.4023	91.69
200	20	5.6133	98.99
230	23	5.8242	106.57
265	26.5	6.1406	118.46
300	30	6.3516	126.74
350	35	6.6680	139.68
400	40	6.9844	153.25
450	45	7.3008	167.45
500	50	7.5117	177.27
550	55	7.8281	192.51
600	60	8.0391	203.03
700	70	8.4609	224.90
800	80	8.7773	242.03
900	90	9.0938	259.80
1000	100	9.4102	278.19
1200	120	10.0430	316.86
1400	140	10.5700	351.01
1600	160	10.9920	379.59
1800	180	11.4140	409.29
2000	200	11.8360	440.10
2300	230	12.4690	488.42
2650	265	13.1020	539.25
3000	300	13.7340	592.60
3500	350	14.4730	658.03

References

1. The European Parliament and the Council of the European Union. REGULATION (EU) 2021/1119 of the European Parliament and the Council of 30 June 2021: Establishing the framework for achieving climate neutrality and amending Regulations (EC) No 401/2009 and (EU) 2018/1999 ('European Climate Law'). *Off. J. Eur. Union* **2021**, *243*, 1–17.
2. Carruth, M.A.; Allwood, J.M.; Milford, R.L. Reducing CO_2 Emissions through Lightweight Design and Manufacturing. In *Reducing CO_2 Emissions through Lightweight Design and Manufacturing, Proceedings of the 14th International Esaform Conference On Material Forming: Esaform 2011, Belfast, UK, 27–29 April 2011*; American Institute of Physics: College Park, MD, USA, 2011; pp. 1632–1637.
3. Huang, R.; Riddle, M.; Graziano, D.; Warren, J.; Das, S.; Nimbalkar, S.; Cresko, J.; Masanet, E. Energy and emissions saving potential of additive manufacturing: The case of lightweight aircraft components. *J. Clean. Prod.* **2016**, *135*, 1559–1570. [CrossRef]
4. Emmelmann, C.; Petersen, M.; Kranz, J.; Wycisk, E. Bionic lightweight design by laser additive manufacturing (LAM) for aircraft industry. In *SPIE Eco-Photonics 2011: Sustainable Design, Manufacturing, and Engineering Workforce Education for a Green Future, Proceedings of the SPIE Eco-Photonics, Strasbourg, France, 28 March 2011*; Ambs, P., Curticapean, D., Emmelmann, C., Knapp, W., Kuznicki, Z.T., Meyrueis, P.P., Eds.; SPIE: Bellingham, WA, USA, 2011; Volume 80650L.
5. Ma, J.; Chen, W.; Zhao, L.; Zhao, D. Elastic Buckling of Bionic Cylindrical Shells Based on Bamboo. *J. Bionic Eng.* **2008**, *5*, 231–238. [CrossRef]
6. Röver, T.; Lau, R.J.; Lange, F.; Struve, A.; Fuchs, C.; Bartsch, K.; Seibel, A.; Emmelmann, C. Methodology for Integrating Biomimetic Beams in Abstracted Topology Optimization Results. In *Volume 4: Biomedical and Biotechnology; Design, Systems, and Complexity, Proceedings of the ASME 2022 International Mechanical Engineering Congress and Exposition, Columbus, OH, USA, 30 October–3 November 2022*; ASME: New York, NY, USA, 2022. [CrossRef]
7. Röver, T.; Bader, M.; Asami, K.; Emmelmann, C.; Kelbassa, I. Development and assessment of a methodology for abstraction of topology optimization results to enable the substitution of optimized beams. *J. Laser Appl.* **2023**, *35*, 042061. [CrossRef]
8. Gralow, M.; Weigand, F.; Herzog, D.; Wischeropp, T.; Emmelmann, C. Biomimetic design and laser additive manufacturing—A perfect symbiosis? *J. Laser Appl.* **2020**, *32*, 21201. [CrossRef]
9. Karam, G.N.; Gibson, L.J. Elastic buckling of cylindrical shells with elastic cores—I. Analysis. *Int. J. Solids Struct.* **1995**, *32*, 1259–1283. [CrossRef]
10. Nachtigall, W.; Wisser, A. *Bionics by Examples: 250 Scenarios from Classical to Modern Times*, 1st ed.; Springer: Berlin/Heidelberg, Germany, 2014; ISBN 978-3-319-05857-3.
11. Pleasant, D.; Gavin, C.; Redden, G.; Nagel, J.; Zhang, H. Bioinspired Design of Material Architecture for Additive Manufacturing. *Machines* **2023**, *11*, 1081. [CrossRef]
12. Mora, A.; Palancares-Díaz, J.; Varela-Soriano, J.; Jimenez-Martinez, M.; Roman-Flores, A.; Farfan-Cabrera, L.I.; Cuan-Urquizo, E. Computational study on the torsional properties of bioinspired tubular metamaterials. *Mech. Adv. Mater. Struct.* **2023**, 1–10. [CrossRef]
13. Ashok, D.; Bahubalendruni, M.V.A.R. Design and Characterization of 2.5D Nature-Inspired Infill Structures under Out-Plane Quasi-Static Loading Condition. *Adv. Mater. Sci. Eng.* **2023**, *2023*, 8918937. [CrossRef]
14. Nachtigall, W.; Wisser, A. *Biologisches Design*; Springer: Berlin/Heidelberg, Germany, 2005; ISBN 3-540-22789-X.
15. Foxtail_1. Banana 1. Licensed under CC BY-NC-SA 2.0. Available online: https://openverse.org/image/4a31b903-ab5b-4485-9a97-3f1b4d65028c/ (accessed on 29 January 2024).
16. SLM Solutions Group AG. SLM®500 Brochure: Production Ready Selective Laser Melting. Available online: https://www.slm-solutions.com/fileadmin/Content/Case_Studies/Production-Ready-Selective-Laser-Melting-SLM500.pdf (accessed on 6 October 2022).
17. Jiang, J.; Xu, X.; Stringer, J. Support Structures for Additive Manufacturing: A Review. *JMMP* **2018**, *2*, 64. [CrossRef]
18. Atzeni, E.; Salmi, A. Study on unsupported overhangs of AlSi10Mg parts processed by Direct Metal Laser Sintering (DMLS). *J. Manuf. Process.* **2015**, *20*, 500–506. [CrossRef]
19. Chen, S.; Tan, Q.; Gao, W.; Wu, G.; Fan, J.; Feng, Z.; Huang, T.; Godfrey, A.W.; Zhang, M.; Huang, X. Effect of heat treatment on the anisotropy in mechanical properties of selective laser melted AlSi10Mg. *Mater. Sci. Eng. A* **2022**, *858*, 144130. [CrossRef]
20. SLM Solutions Group AG. Material Data Sheet Al-Alloy AlSi10Mg. Available online: https://www.slm-solutions.com/fileadmin/Content/Powder/MDS/MDS_Al-Alloy_AlSi10Mg_0520_EN.pdf (accessed on 24 August 2022).
21. Sert, E.; Schuch, E.; Öchsner, A.; Hitzler, L.; Werner, E.; Merkel, M. Tensile strength performance with determination of the Poisson's ratio of additively manufactured AlSi10Mg samples. *Materialwiss. Werkstofftech.* **2019**, *50*, 539–545. [CrossRef]
22. COMSOL. Reference Manual: Version: Comsol Multiphysics 6.0. Available online: https://doc.comsol.com/6.0/doc/com.comsol.help.comsol/COMSOL_ReferenceManual.pdf (accessed on 4 April 2023).
23. COMSOL. Optimization Module User's Guide: Version: Comsol Multiphysics 6.0. Available online: https://doc.comsol.com/6.0/doc/com.comsol.help.opt/OptimizationModuleUsersGuide.pdf (accessed on 4 April 2023).
24. Han, Q.; Gu, H.; Soe, S.; Setchi, R.; Lacan, F.; Hill, J. Manufacturability of AlSi10Mg overhang structures fabricated by laser powder bed fusion. *Mater. Des.* **2018**, *160*, 1080–1095. [CrossRef]
25. Nelder, J.A.; Mead, R. A Simplex Method for Function Minimization. *Comput. J.* **1965**, *7*, 308–313. [CrossRef]
26. Nelder, J.A.; Mead, R. Errata. *Comput. J.* **1965**, *8*, 27. [CrossRef]
27. Nocedal, J.; Wright, S.J. *Numerical Optimization*, 2nd ed.; Springer: New York, NY, USA, 2006; ISBN 0-387-30303-0.

28. Öchsner, A. *Classical Beam Theories of Structural Mechanics*; Springer International Publishing: Cham, Switzerland, 2021; ISBN 978-3-030-76034-2.
29. Autodesk Inc. Autodesk Netfabb 2023: Product Website. Available online: https://help.autodesk.com/view/NETF/2023/ENU/ (accessed on 26 February 2024).
30. One Click Metal GmbH. Technical Specifications—MPRINT. Available online: https://oneclickmetal.com/boldseries-technical-specifications/ (accessed on 26 February 2024).
31. Röver, T.; Fuchs, C.; Asami, M.K.; Emmelmann, C. Parameter Optimization Files (COMSOL Multiphysics 6.0) for Dimensioning of Biomimetic Beams under Bending. 2024. Available online: https://doi.org/10.15480/882.9113 (accessed on 3 March 2024).

A.4 Novel lightweight design of additively manufactured structural components through biomimetics

The following pages feature a reprint of the journal article "Novel lightweight design of additively manufactured structural components through biomimetics," originally published in 2025. This reprint is provided with permission from the Japan Society of Powder and Powder Metallurgy (JSPM) granted through the Japan Academic Association for Copyright Clearance (JAC).

J. Jpn. Soc. Powder Powder Metallurgy, **72** (2025) S1241-S1248
https://doi.org/10.2497/jjspm.16C-T11-10

Novel Lightweight Design of Additively Manufactured Structural Components through Biomimetics

Tim RÖVER[1*], Tjark SIEVERS[1], Karim ASAMI[1], Claus EMMELMANN[1]

[1]Hamburg University of Technology, Institute of Laser and Systems Technology, Harburger Schloßstraße 28, 21079 Hamburg, Germany

Abstract

PBF-LB/M offers efficient production of complex and functional metal components. Research shows that the mechanical performance of components can be enhanced by incorporating biomimetic shapes, such as banana pseudo stem-inspired biomimetic beams. This work addresses three mechanical topology optimization problems, using AlSi10Mg. Leveraging these models, novel component designs featuring biomimetic beams and novel node designs were developed based on the input topology optimization. The new biomimetic component designs were compared to their topology optimization counterparts that were used as references. Numerical analyses of mass and structural integrity were performed to evaluate improvements. Lightweight biomimetic component designs for additively manufactured components were developed using the developed methods. Weight savings of 12.5-30.3 % were achieved compared to the input topology optimization results. However, further research on the design methodologies is needed to ensure that the mechanical stress criterion is also met within the nodes of the generated biomimetic designs.

Keywords: Design for additive manufacturing (DfAM), powder bed fusion of metal with laser beam (PBF-LB/M), aluminum alloy AlSi10Mg, biomimetics, lightweight design

1. Introduction

The combination of laser additive manufacturing and biomimetics resulted in numerous efficient component designs [1–3]. However, the use of biomimetics in Additive Manufacturing (AM) is still limited. One of the limiting factors are missing design tools [1]. This study presents a detailed methodology for biomimetic design of mechanical components from topology optimization (TO) results to partly close this gap in research. A study compared the conventional design of an aircraft bracket with a newly developed design of the bracket that combined additive manufacturing and biomimetics [4]. An initial design of the bracket was generated by TO. In the next step, a bamboo-inspired structure was introduced to improve the parts lightweight character. The authors state, that a change of the material, manufacturing technique and design resulted in a weight reduction of 50 % when comparing the newly developed part with the original part [4]. Based on this approach a methodology for integration of biomimetic beams in TO results was developed [5]. The methodology focuses on mechanical TO results that are truss-like and consist of members and nodes. The authors suggest five conceptual design steps on how to advance such a TO result to a biomimetic component design [5]. Building on this approach a detailed abstraction methodology [6] and a design tool for biomimetic beams under bending were presented in other studies. A research gap is given by a missing detailed solution for node design that is in line with the method from Röver et al. [5] and application of the resulting full design process to exemplary TO results. Therefore, the goal of this paper is to close this research gap.

2. Materials and Methods

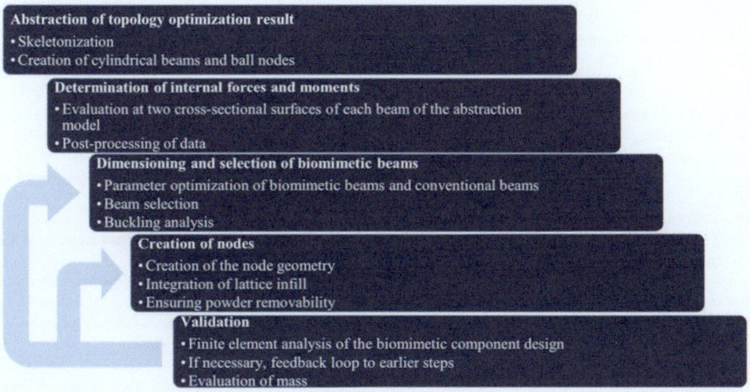

Fig. 1. Flowchart of methodology for development of additively manufactured structural components through biomimetics based on topology optimization results.

*corresponding author, E-mail: tim.roever@tuhh.de

Fig. 1 shows the flowchart of the design methodology that was developed and applied within this work. It is based on the previous study [5] and focuses on the powder bed fusion of metal with laser beam (PBF-LB/M) process as the AM process of choice due to its great relevance in industry. The proposed methodology involves a comprehensive five-step approach to optimize and validate biomimetic structures that are developed from TO results. The first step, abstraction of TO results, employs skeletonization to derive a simplified model consisting of cylindrical beams and ball nodes. This abstraction facilitates the determination of internal forces and moments in the second step, where evaluations are conducted at two cross-sectional surfaces of each beam, followed by data post-processing. The third step focuses on dimensioning and selecting biomimetic beams. This includes parameter optimization for both biomimetic and conventional beams, beam selection, and conducting a buckling analysis. The creation of nodes in the fourth step involves developing node geometries, integrating lattice infills, and ensuring good powder removability, which is crucial for the PBF-LB/M process. Finally, the validation step encompasses Finite Element Analysis (FEA) of the biomimetic component design to verify structural integrity. If the stress criterion (acceptance criterion) is not satisfied, a feedback loop allows for iterative refinement of earlier steps. Additionally, the evaluation of mass allows comparison of the biomimetic design with the original TO design. Heat-treated AlSi10Mg was considered as a material in this study. Material properties can be found in Table 1 and are based on literature [7–9]. Comsol Mutliphysics 6.0 [10], Autodesk Inventor 2024 [11] and nTop 4.22.2 [12] are the software that were used in this work.

Table 1 Material properties of heat-treated AlSi10Mg manufactured by PBF-LB/M. Adapted from literature [7] (licensed under CC BY 4.0).

Material Property	Value
Yield strength	141 MPa [8]
Young's modulus	57 GPa [8]
Poisson's ratio	0.34 [9]
Density	2.67 g/cm^3 [8]

3. Results

In this section the results of the study are presented.

A. Reference Topology Optimization Results and Abstractions

The TO results and abstractions used in this work were taken from a previous study [6] and its corresponding data set [13]. The study [6] considered structural steel as material for all components. In contrast, aluminum alloy AlSi10Mg (see Table 1) was considered for all components in this study. Due to the manufacturing restrictions of the PBF-LB/M process such as minimum wall thicknesses and minimum distances between opposing walls, it was assumed that larger components would have a greater potential for improvement in terms of lightweighting. Therefore, the TO results and abstractions were scaled up. The size of the respective TO results and abstractions was increased by factors of 4, 4, and 1.5 for the Michell cantilever, the Messerschmitt-Bölkow-Blohm (MBB) beam and the L-shape, respectively (see Fig. 7 (a)-(c)). Certain PBF-LB/M systems have relatively large build envelopes with dimensions of e.g. 500 mm × 280 mm × 850 mm [14]. Therefore, all upscaled designs fit into the build envelope of such PBF-LB/M systems. Furthermore, the abstraction of the Mitchell cantilever was complemented by adding two beams (see #15 and #16 in Fig. 7 (f)) that connect the abstraction with the mechanical support.

Based on the changes, the mechanical loads of the upscaled TO designs had to be increased to allow for a meaningful comparison to the biomimetic designs later on in the study (see Table 3). For identification of suitable forces, the TO results (see Fig. 7 (a)-(c)) were considered. For each result, the force that would lead to a maximum von Mises stress of 80 % inside the component, was identified by a numerical parameter optimization study in COMSOL Multiphysics 6.0. It is noted that stresses were not evaluated close to the perfectly stiff supports as well as close to the surfaces to which the forces were applied. This approach prevents evaluating physically meaningless stress peaks that are caused by discretization of the mechanical problem. The identified loads can be found in Table 2.

Table 2 Geometric scaling factors and identified loads for design study

Model	Geometric scaling factor	Load [N]
Michell cantilever	4	8237.5
MBB beam	4	16044
L-shape	1.5	2981

B. Evaluation of Internal Forces and Moments

For evaluation of internal forces and moments of the abstractions (see Fig. 7 (d)-(f)), FEAs were performed in COMSOL Multiphysics 6.0. The upscaled mechanical boundary conditions were considered. Furthermore, for each beam, work planes were added if possible at 20 % of the beam length and at 80 % of the beam length according to the study by Röver et al. [5]. If two beams intersected at an acute angle, a work plane might intersect with a neighboring beam. In such cases the work plane was moved towards the center of the beams away from the intersection of beams (see Fig. 2) to prevent that evaluation planes would intersect with neighboring beams. According to literature [6], the section forces function as implemented in COMSOL Multiphysics 6.0 was used for all evaluation planes. Doing so for each evaluation plane the axial force N, the shear force along the first local axis T_1, the shear force along the second local axis T_2, the twisting moment M_t, the bending moment around the first local axis M_1, and the bending moment around the second local axis M_2 could be identified.

According to literature it can be assumed that N, T_1, T_2, and M_t are constant along the length of the beams [5]. Due to numerical discretization of the mechanical problem usually small deviations are present when comparing the numerical values that were evaluated at the planes at 20 % of the length and at 80 % of the length of the beams. In these cases, the average of the two values was used as an approximation for the next steps of the methodology.

M_1 and M_2 cannot be assumed to be constant along the length of the beams. The bending moments within each beam occur from the shear forces and the bending moments applied to the ends of the beams. It can be assumed that the bending moments that occur at the cross-sections of the beams change linearly with respect to the beam length [5]. Using the bending moments values evaluated at each of the two planes of a beam, the bending moment values are extrapolated to 0 % and 100 % of the length of the respective beam using a spreadsheet. This approach is in line with one of the previous studies on this method [5].

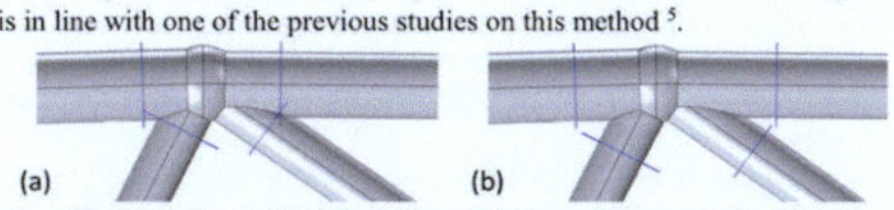

Fig. 2. Exemplary critical node with planes for evaluation of internal forces and moments in abstractions. (a) planes for evaluation at 20 % and 80 % of lengths of beams that intersect with planes of neighboring beams; (b) alternative and preferred locations of planes for evaluation.

C. Dimensioning and Selection of Biomimetic Beams

Dimensioning and selection of beams in this work was based on the study by Röver et al. [7] and the respective dataset [15]. The work provides parametric designs of beams and parameter optimization models for dimensioning of the beams under bending loads. Fig. 3 (a), (c) and (d) show the parametric cross-section designs of solid cylindrical beam (SC), revolver drum beam 1 (RD1) and revolver drum beam 2 (RD2) according to the study, respectively. It is noted, that the design parameter intervals as provided by the authors of the study, consider manufacturing restrictions of the PFB-LB/M process. The three previously existing designs were augmented in this study by a parametric design of a hollow cylindrical beam (HC) that is shown in Fig. 3 (b). Based on the parametric HC design a corresponding parameter optimization model of the HC was developed that is analogous to the existing models with respect to the finite element mesh, the optimization settings and the mechanical constraints.

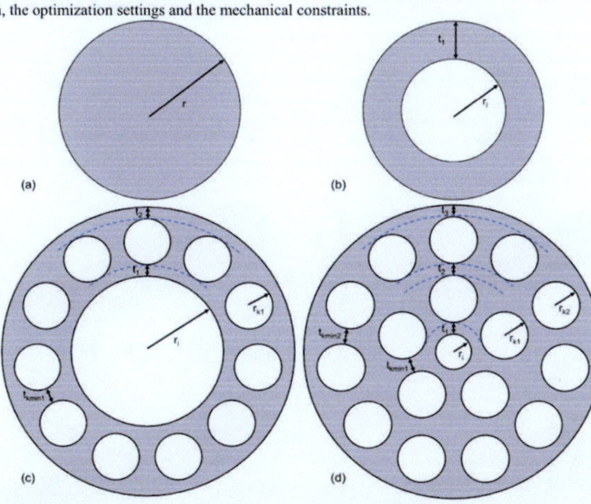

Fig. 3. Parametric beam designs. (a) solid cylindrical beam (SC); (b) hollow cylindrical beam (HC); (c) revolver drum beam 1 (RD1); (d) revolver drum beam 2 (RD2). Adapted and partly reprinted from literature [7] (licensed under CC BY 4.0).

The models from the work by Röver et. al. [7] are fixed to a length of 100 mm and considered a shear force resulting in a bending moment only. They are mechanically fixed at one end and have a boundary force acting at their free end. The optimization models minimize the cross-sectional area of the beams while considering certain constraints.

The parametric models of SC, HC, RD1 and RD2 were improved in this paper. The meshing sequence was improved such that the number of elements increases with respect to the length of the beam. Based on the work by Röver et. al. [7] the element length in direction of the main axis of the beam was set to approximately one millimeter. Furthermore, to achieve more accurate numerical results, the element length in direction of the main axis of the beam was halved for the elements close to the mechanical support as well as the elements close to the boundary where the forces and moments are applied. Additionally, the models were augmented by using the rigid connector function in COMSOL Multiphysics 6.0, to efficiently apply forces and moments to the free ends of the beams. The parameter optimization models in this work allow for application of an axial force, a shear force along the first local axis, a shear force along the second local axis, a twisting moment, a bending moment around the first local axis, and a bending moment around the second local axis.

The upper limits for occurring von Mises stresses in the domains of interest of all parameter optimization models were modified to 80 % of the yield strength of AlSi10Mg, which is 112.8 MPa (see also Table 1). The RD2 design has a minimum outer diameter of 21 mm considering the defined parameter intervals. Therefore, to reduce computational efforts within the study, RD2 optimizations were only run for a beam, if the result of the RD1 optimization gave a beam with an outer diameter larger than 21 mm.

Fig. 4 shows optimization results of all four types of beams based on the evaluated forces and moments and the length of one exemplary cylindrical beam of the abstraction of a TO result. Therefore, all four beams are subject to the same mechanical boundary conditions and have the same length. It can be seen that regions of high stress in all results are close to the maximum allowable stress of 112.8 MPa. In the next step, the beam that has the lowest cross-sectional area is chosen. This value can be evaluated in the optimization models with low effort and compared in a spreadsheet.

To ensure structural integrity of the selected beams, beams that are under compression have to be analyzed with regard to buckling. For this step, buckling analysis is done in COMSOL Multiphysics 6.0 using the same material parameters, the same mechanical boundary conditions and the same meshing sequence as was used for the final result of the corresponding parameter optimization. By evaluation of the critical load factors, it can be estimated whether or not the beams might be unstable with regard to buckling. If the lightest of the four beams is uncritical with regard to buckling, it is chosen. If a beam is found to be critical with regard to buckling, the following iterations are performed: Increase of the wall thickness of the outermost ring (parameter r of SC design, parameter t_1 for HC design, parameter t_2 for RD1 design and parameter t_3 for RD2 design) by small increments. Identification of a small value of the respective parameter for that the beam design is not critical with regard to buckling according to buckling analysis. Doing so, one valid solution for the respective beam can be found. The same may be done for the other three beams. Then again, the beam with the lowest cross-sectional area is chosen for the next steps of the methodology. It is noted that not all four beams have to be investigated that way, if one initial solution of a beam is uncritical to buckling. In this case, beams with higher cross-sectional areas do not need to be investigated in this way. Alternatively, if a SC beam is found to be critical with regard to buckling, dimensioning may be done using the Euler buckling criterion considering a fixed-free condition according to literature [16,17].

After all final beam designs were selected, the beams are introduced into a new component in COMSOL Multiphysics 6.0. The beams are oriented based on the orientation of beams in the abstraction of the TO. Furthermore, it has to be ensured that their orientation (rotation about main axis of beam) is in line with the optimization simulations, as RD1 and RD2 designs are not fully rotationally symmetric. Additionally, sheet-like elements are added in regions where supports are located and in regions were loads are applied in the original TOs. In preparation of the step of node generation, all beams are shortened such that the beams do not intersect and additionally have a certain distance between them that allows for a smooth node. The outcome of this step for the three components investigated in this work can be seen in Fig. 7 (g)-(i). These preliminary component designs are exported in Standard for the Exchange of Product model data (STEP)-file format.

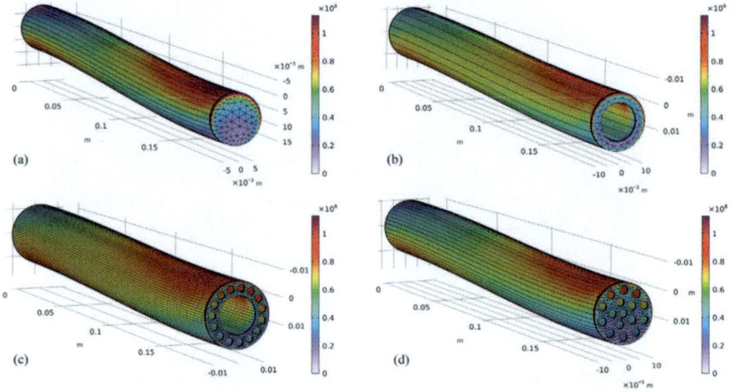

Fig. 4. Exemplary finite element method-based parameter optimization results of beams subject to the same mechanical boundary conditions. Deformation of the beams are exaggerated for improved visibility. (a) solid cylindrical beam (SC); (b) hollow cylindrical beam (HC); (c) revolver drum beam 1 (RD1); (d) revolver drum beam 2 (RD2).

D. Generation of Nodes

Röver et al. suggest TO for generation of lightweight node designs [5]. As generation of high-quality TO results often requires a lot of time, effort, and, experienced engineers are needed, a different approach was followed in this study. For this study, the following goals were defined for development of a methodology for node design between the optimized beams: 1. Transmission of forces and moments between beams, 2. Ensure structural integrity, 3. Ensure powder removability from node as well as cavities of beams connected to node, 4. Ensure that node design and resulting final component design consider manufacturing restrictions of PBF-LB/M, 5. Small mass of node, and 6. Ensure that design sequence of nodes is robust and time-efficient. One promising approach for node design was found to be the use of lattice structures. Certain lattice structures were found to be well suited with regards to powder removability and were found to show high potential for lightweight design [18]. Furthermore, many lattice structures are self-supporting in PBF-LB/M. Node generation was performed in two steps: the generation of solid volume nodes and the integration of lattice infill structure into the nodes.

For generation of the solid volume nodes Autodesk Inventor 2024 was used. The corresponding design steps based on an exemplary node are depicted in Fig. 5. First the STEP-file containing the optimized beams and sheet-like elements is imported into the software.

Then, the following steps are carried out for all nodes: First, a 2 mm offset is added to the outer diameter of the end of the beam that is facing towards the node. This is done for all beams of the node (see Fig. 5 (a)). In the next step, the loft function is used to connect the offset circles of all beams of a node with each other (see Fig. 5 (b), (c)&(d)). After using the loft function, a node may have one or more holes near its center. If so, the sketch function and extrusion function are used to generate a solid volume at the center of the nodes, closing any holes. The extrusion thickness is based on the outer radii of the beams connected to the node. Next, the fillet function is used for smoothing of the node (see Fig. 5 (e)&(f)). Smoothing is also done in direction of the center parts of the beams such that additional material is added to the beams. Then the node geometry is separated from the beams by cuts using work planes (see Fig. 5 (g)&(h)). To achieve the final solid volume node, extruded cuts of the cavities of the beams into the node are done (see Fig. 5 (i)). The length of the extruded cuts is suggested to be equal or greater than the later used unit cell size of the lattice structure. Then a single STEP-file is exported, that contains the whole component geometry including beams and nodes. The beams and nodes are defined as solid volumes within the STEP-file.

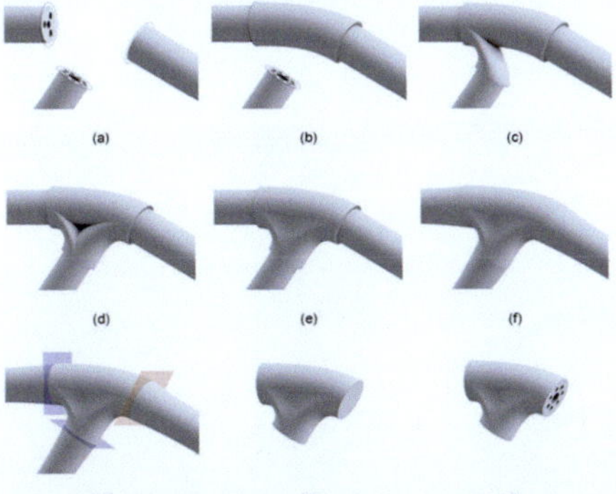

Fig. 5. Generation of an exemplary solid volume node in five steps. (a) added offsets; (b), (c)&(d) connection of all beams; (e)&(f) smoothing; (g)&(h) separation of node; (i) extrusion of cavities into nodes.

For integration of the lattice infill structures into the nodes and finalization of the node design, nTop 4.22.2 is used. First, the STEP-file is imported to nTop. Next, the nodes and beams of the design are converted to implicit bodies. Then the following steps are carried-out for all nodes (see Fig. 6): First a shell is generated from the solid volume (see Fig. 6 (a)) and perforations are added to further modify the shell (see Fig. 6 (b)). Shell wall thicknesses of 3 mm provided good results for many nodes generated in this study. Next, circular cut-outs are added for powder removal to achieve the final shell of the node (see Fig. 6 (c)). Next, the solid volume of the initial node is filled with a Kelvin cell to generate a lattice structure (see Fig. 6 (d)&(e)). The design parameters of the Kelvin cell should be chosen in such a way that powder can be removed through the structure and manufacturing constraints of PBF-LB/M are considered. In this work Kelvin cells with unit cell size of 5 mm × 5 mm × 5 mm and a thickness value between 2 mm and 2.5 mm were used. In the final step, the shell of the node and its lattice infill are combined to obtain the final node design (see Fig. 6 (f)&(g)). By application of this method lattice-filled nodes can be generated.

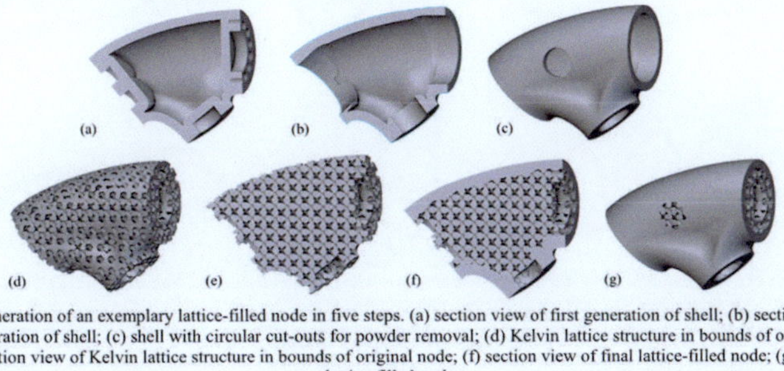

Fig. 6. Generation of an exemplary lattice-filled node in five steps. (a) section view of first generation of shell; (b) section view of second generation of shell; (c) shell with circular cut-outs for powder removal; (d) Kelvin lattice structure in bounds of original node; (e) section view of Kelvin lattice structure in bounds of original node; (f) section view of final lattice-filled node; (g) final lattice-filled node.

E. Complete Biomimetic Designs

The modification of nodes and introduction of infill structures according to the previous section can be accomplished in nTop, thereby enabling the generation of a complete component design that incorporates biomimetic beams and lattice-filled nodes. It is noted that the generation of the complete biomimetic designs of the three components exhibits certain unique aspects, which are outlined in the following paragraphs.

In the beam optimizations of all three component designs, selected parameters were restricted using additional upper bounds to reduce the computational effort. For SC parameter r was additionally restricted, for HC parameter t_1 was additionally restricted, for RD1 t_1 and t_2 were additionally restricted and for RD2 t_1, t_2 and t_3 were additionally restricted. This was necessary to achieve optimization results within an acceptable time frame. The final optimization results show that the values of the respective design parameters are considerably lower than the chosen upper bounds for all cases. Therefore, it can be expected that the additional restriction of the solution space had no negative impact on the optimization results.

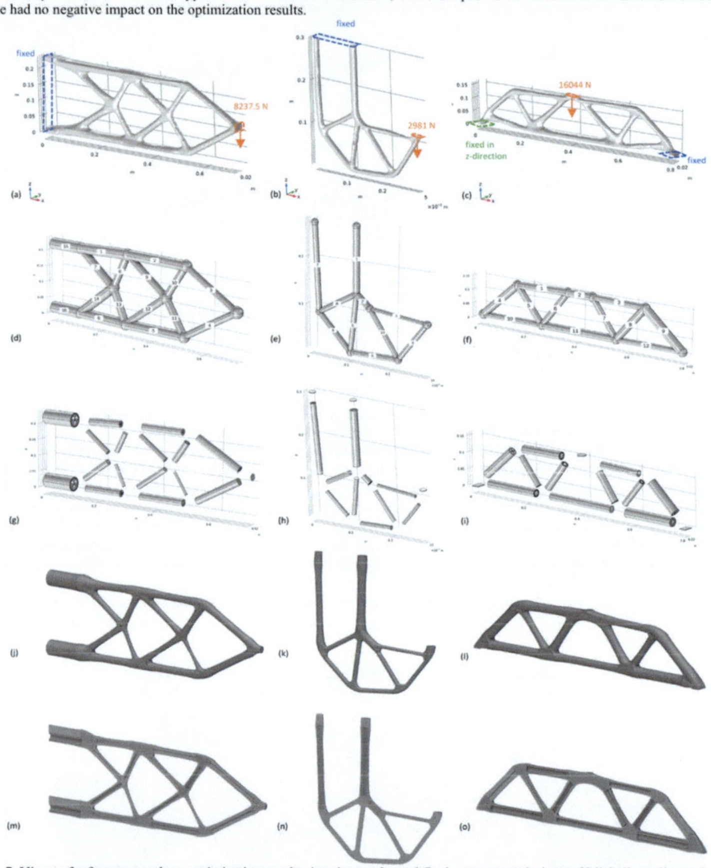

Fig. 7. Views of reference topology optimization results, interim results and final component designs of Michell cantilever, L-shape and MBB beam. (a)-(c): topology optimization results with boundary conditions. (d)-(f): abstractions of topology optimization results. (g)-(i): arrangements of biomimetic beams. (j)-(l): final biomimetic component designs. (m)-(o): section views of final biomimetic component designs.

The L-shape contains seven SC beams. As SC beams do not have cavities, removal of internal powder is not necessary. The nodes linking SC beams were generated using a modified design approach. During generation of the solid volume nodes, no 2 mm offsets were added to the outer diameter of the end of SC beams. Furthermore, if a node connected only SC beams no lattice infill was used, but solid volume nodes were generated. For nodes that connect SC beams with other kinds of beams, a lattice structure was used in those areas of the node that connect beams with cavities. Regions of the nodes close to SC beams were designed as solid volumes if lattice structure was not necessary for powder removal. This modified design approach was used to ensure better powder removability and provide more material in the nodes close to the SC beams that are solid themselves for an increased structural integrity. Furthermore, the L-shape was the only component of the three components in which beams under compressive loads were found to be critical with regard to buckling. All three critical beams were SC beams. According to the presented methodologies the Euler buckling criterion was used for dimensioning of SC beams. Subsequent buckling analysis of the newly dimensioned SC beams yielded no risk for buckling. Additionally, when comparing the updated beams with the other beams of the same load case, they were still the lightest and, therefore, used for the final component design.

During application of the design methodology, beam #2 of the MBB beam (see Fig. 7 (f)) was omitted as it was relatively short. Also, the sheet element where the force is applied, was integrated into the node.

In the implementation of the design methodology for the Michell cantilever, one node was designed on an individual basis, diverging from the presented design methods. For this node that connects one SC beam and three HC beams, an alternative node design without a lattice structure was employed. A solid volume node was generated and openings from the HC cavities to the exterior were added with the loft function in Autodesk Inventor 2024. Moreover, in the case of the Michell cantilever, the SC, the HC and the RD2 beam optimizations were conducted for selected beam load cases, only, thus reducing the computational effort in this part of the study. Furthermore, RD2 was preselected and optimized for beams #15 and #16 to reduce the computational effort.

The complete biomimetic component designs can be found in Fig. 7 (j)-(o). The biomimetic Michell cantilever features all four types of beams, whereas the biomimetic MBB beam only features HC and RD1 beams. The biomimetic L-shape mostly utilizes SC beams and three RD1 beams.

F. Finite Element Analysis of the Biomimetic Component Designs

FEAs of the three biomimetic component designs were conducted. A component design would be considered mechanically acceptable (acceptance criterion), if the von Mises stresses did not exceed the yield strength of the material (141 MPa). At the same time good comparability to the TO designs would be given if the von Mises stress did not exceed 80 % of the yield strength of the material (112.8 MPa) as this threshold was used for scaling of the force values with the scaled TO results.

Fig. 8 shows a section view of the biomimetic MBB beam. It can be seen that in multiple nodes in the finite element mesh 80 % of the yield strength of the material is exceeded (these nodes are highlighted in blue). It can be seen that the critical areas are rather small and are located at the unions of the nodes and the beams. Additionally, a small number of critical nodes were located in the central parts of certain nodes. Variation of the node design parameters including the choice of the type of lattice structure was done. However, no design could be generated which satisfied the acceptance criterion. Also, the biomimetic Michell cantilever showed areas in which the yield strength of the material was exceeded. Critical regions were located at the unions of the nodes and the beams, regions inside the nodes including the lattice structures of the nodes, regions close to supports and regions close to where forces were applied. In the case of the L-shape, the yield strength of the material was exceeded at unions of the nodes and the beams as well as certain regions inside the nodes.

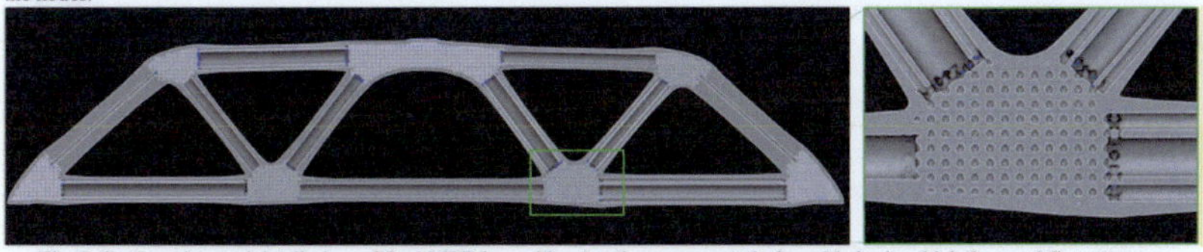

Fig. 8. Section view and detail view of final MBB beam biomimetic component designs. Nodes in which the von Mises stress exceeds 80 % of the yield strength of the material (112.8 MPa) are highlighted in blue.

Furthermore, the yield strength of the material was exceeded in regions close to supports and regions close to where forces are applied. Also, for the Michell cantilever and the L-shape variation of the node design parameters was done. However, also for these components no designs could be generated which satisfied the acceptance criterion. Thus, all three biomimetic designs could not satisfy the acceptance criterion based on the yield strength of the material. At the same time, it could be shown that the biomimetic beams in all design were well dimensioned based on the utilized methodology.

Table 3 Comparison of mass of topology optimization designs and biomimetic component designs

Component design	Mass [g]	Weight reduction compared to topology optimization result
Michell cantilever – topology optimization	3069.3	-
Michell cantilever - Biomimetic	2139.7	30.3 %
MBB beam - topology optimization	2220.4	-
MBB beam - Biomimetic	1889	14.9 %
L-shape - topology optimization	384.6	-
L-shape - Biomimetic	336.5	12.5 %

G. Assessment of the Lightweight Performance

For assessment of the lightweight performance of the biomimetic component designs, their masses were evaluated. Furthermore, the mass of each of the TO results was evaluated. Comparison of biomimetic designs with original TO results yielded a weight reduction of 30.3 %, 14.9 % and 12.5 % for Michell cantilever, MBB-beam and L-shape, respectively. Results of evaluation are summarized in Table 3.

4. Discussion

The research gap addressed by this study could be partly closed as a detailed solution for node design was developed that is in line with the method from [5]. Furthermore, an application of the full design process to three exemplary TO results was done such that three biomimetic component designs were generated. Improvement of a design that was generated by well-performed TO is a challenging task, as such a result is a mathematical optimum. Therefore, the weight reductions of 12.5-30.3 % compared to the original TO results that were achieved in this study show the great potential of the utilized approach and the developed methods. The design methods show high potential for a fast and cost-efficient design of lightweight components that are produced by PBF-LB/M and may contribute to a more sustainable mobility in sectors such as automotive and aerospace. FEA of the biomimetic designs yielded that von Mises stresses exceeded the yield strength of the material in certain nodes in all of the three components. Efforts to find alternative design solutions for the nodes have so far been unsuccessful. Therefore, the proposed design methodology could not be validated successfully. Even though the acceptance criterion could not be satisfied, the developed designs show a high potential as only the nodes in the biomimetic component designs were critical in terms of the occurring von Mises stresses. The authors assume that the design methodology could be improved to achieve better node designs. A focus should be put on an improved transition between node designs and beam designs as well as achieve a lattice structure that prevents peak stresses more efficiently. This may achieve overall component designs that satisfy the stress criterion. At the same time, the authors of this study believe that it is possible to achieve mechanically improved nodes that have approximately the same weight as the current designs but satisfy the stress criterion. Therefore, this study contributes to closing the addressed research gap and presents interim results on node design for biomimetic components as well as identifies node design as one aspect that should be addressed in future works.

5. Conclusions and Future Prospects

This study focused on the advancement, application and assessment of design methodologies for additively manufactured biomimetic structural components. A design methodology for node design of biomimetic structural components was developed. The methodology builds on previous works in this field. Furthermore, the previously existing design methodologies and the newly developed methodology were combined and applied to three TO results to achieve novel biomimetic structural components. FEA showed that in all biomimetic component designs, stresses above the yield strength of the material occurred. Therefore, the design criterion could not be satisfied. At the same time comparison of biomimetic designs with the original TO results yielded a weight reduction of between 12.5 and 30.3 %. Therefore, valuable design solutions and the great potential of the design approach could be demonstrated in this study. Future research should aim to enhance the node design methodology in order to develop an overarching design methodology that can satisfy the stress criterion during the application of the aforementioned methodologies while achieving low component masses.

References

1. Gralow, M., Weigand, F., Herzog, D., Wischeropp, T. & Emmelmann, C. Biomimetic design and laser additive manufacturing—A perfect symbiosis? *Journal of Laser Applications* **32**, 021201 (2020).
2. Du Plessis, A. *et al.* Beautiful and Functional: A Review of Biomimetic Design in Additive Manufacturing. *Additive Manufacturing* **27**, 408–427 (2019).
3. Yang, Y. *et al.* Recent Progress in Biomimetic Additive Manufacturing Technology: From Materials to Functional Structures. *Advanced Materials* **30**, 1706539 (2018).
4. Emmelmann, C., Petersen, M., Kranz, J. & Wycisk, E. Bionic lightweight design by laser additive manufacturing (LAM) for aircraft industry. in *SPIE Eco-Photonics 2011: Sustainable Design, Manufacturing, and Engineering Workforce Education for a Green Future* (eds. Ambs, P. et al.) 80650 (SPIE, 2011). doi:10.1117/12.898525.
5. Röver, T. *et al.* Methodology for Integrating Biomimetic Beams in Abstracted Topology Optimization Results. in *Proceedings of the ASME 2022 International Mechanical Engineering Congress and Exposition. Volume 4: Biomedical and Biotechnology; Design, Systems, and Complexity* (Columbus, Ohio, USA, 2022). doi:10.1115/IMECE2022-94299.
6. Röver, T., Bader, M., Asami, K., Emmelmann, C. & Kelbassa, I. Development and assessment of a methodology for abstraction of topology optimization results to enable the substitution of optimized beams. *Journal of Laser Applications* **35**, (2023).
7. Röver, T., Fuchs, C., Asami, K. & Emmelmann, C. Dimensioning of Biomimetic Beams under Bending for Additively Manufactured Structural Components. *Biomimetics* **9**, 214 (2024).
8. SLM Solutions Group AG. Material Data Sheet Al-Alloy AlSi10Mg.
9. Sert, E. *et al.* Tensile strength performance with determination of the Poisson's ratio of additively manufactured AlSi10Mg samples. *Materialwiss. Werkstofftech.* **50**, 539–545 (2019).
10. COMSOL. COMSOL Multiphysics - Reference Manual - Version: 6.0. (2021).
11. Autodesk Inc. Inventor 2024 Help | Inventor API User's Manual | Autodesk. https://help.autodesk.com/view/INVNTOR/2024/ENU/?guid=UserManualIndex.
12. nTopology Inc. Next-Gen Engineering Design Software: nTop (formerly nTopology). *nTop* https://www.ntop.com/.
13. Röver, T., Bader, M., Asami, M. K., Emmelmann, C. & Kelbassa, I. Supplementary material to article with title: Development and Assessment of a Methodology for Abstraction of Topology Optimization Results to Enable the Substitution of Optimized Beams. https://doi.org/10.15480/882.8168 (2023).
14. SLM Solutions Group AG. SLM®800 Brochure - Selective Laser Melting Machine - Large Format Selective Laser Melting.
15. Röver, T., Fuchs, C., Asami, M. K. & Emmelmann, C. Parameter Optimization Files (COMSOL Multiphysics 6.0) for Dimensioning of Biomimetic Beams under Bending. https://doi.org/10.15480/882.9113 (2024).
16. *Technische Mechanik. 2: Elastostatik / Dietmar Gross, Werner Hauger, Jörg Schröder, Wolfgang A. Wall.* (Springer Vieweg, Berlin; [Heidelberg], 2017). doi:10.1007/978-3-662-53679-7.
17. Euler, L. *Methodus Inveniendi Lineas Curvas Maximi Minimive Proprietate Gaudentes, Sive Solutio Problematis Isoperimetrici Lattissimo Sensu Accepti.* (Lausannæ & Genevæ, 1744).
18. Tao, W. & Leu, M. C. Design of lattice structure for additive manufacturing. in *2016 International Symposium on Flexible Automation (ISFA)* 325–332 (IEEE, Cleveland, OH, USA, 2016). doi:10.1109/ISFA.2016.7790182.

Received July 2, 2024; Revised October 17, 2024; Accepted October 21, 2024

MIX
Papier aus verantwortungsvollen Quellen
Paper from responsible sources
FSC® C105338

If you have any concerns about our products,
you can contact us on
ProductSafety@springernature.com

In case Publisher is established outside the EU,
the EU authorized representative is:
Springer Nature Customer Service Center GmbH
Europaplatz 3, 69115 Heidelberg, Germany

Printed by Libri Plureos GmbH
in Hamburg, Germany